THE PHILOSOPHICAL PROJECT
OF CARNAP AND QUINE

Rudolf Carnap (1891–1970) and W. V. Quine (1908–2000) have long been seen as key figures of analytic philosophy who are opposed to each other, due in no small part to their famed debate over the analytic/synthetic distinction. This volume of new essays assembles for the first time a number of scholars of the history of analytic philosophy who see Carnap and Quine as figures largely sympathetic to each other in their philosophical views.

The essays acknowledge the differences which exist, but through their emphasis on Carnap and Quine's shared assumption about how philosophy should be done – that philosophy should be complementary to and continuous with the natural and mathematical sciences – our understanding of how they diverge is also deepened. This volume reshapes our understanding not only of Carnap and Quine, but of the history of analytic philosophy generally.

SEAN MORRIS is Professor of Philosophy at Metropolitan State University of Denver. His previous publications include *Quine, New Foundations, and the Philosophy of Set Theory* (Cambridge, 2018).

THE PHILOSOPHICAL PROJECT OF CARNAP AND QUINE

EDITED BY

SEAN MORRIS

Metropolitan State University of Denver

CAMBRIDGE
UNIVERSITY PRESS

University Printing House, Cambridge CB2 8BS, United Kingdom

One Liberty Plaza, 20th Floor, New York, NY 10006, USA

477 Williamstown Road, Port Melbourne, VIC 3207, Australia

314–321, 3rd Floor, Plot 3, Splendor Forum, Jasola District Centre,
New Delhi – 110025, India

103 Penang Road, #05–06/07, Visioncrest Commercial, Singapore 238467

Cambridge University Press is part of the University of Cambridge.

It furthers the University's mission by disseminating knowledge in the pursuit of education, learning, and research at the highest international levels of excellence.

www.cambridge.org
Information on this title: www.cambridge.org/9781108494243
DOI: 10.1017/9781108664202

© Cambridge University Press 2023

This publication is in copyright. Subject to statutory exception and to the provisions of relevant collective licensing agreements, no reproduction of any part may take place without the written permission of Cambridge University Press.

First published 2023

A catalogue record for this publication is available from the British Library.

Library of Congress Cataloging-in-Publication Data
NAMES: Morris, Sean (Professor of Philosophy), editor.
TITLE: The philosophical project of Carnap and Quine / edited by Sean Morris, Metropolitan State University of Denver.
DESCRIPTION: Cambridge, United Kingdom ; New York, NY, USA : Cambridge University Press, 2023. | Includes bibliographical references and index.
IDENTIFIERS: LCCN 2022024945 | ISBN 9781108494243 (hardback) |
ISBN 9781108664202 (ebook)
SUBJECTS: LCSH: Analysis (Philosophy) | Carnap, Rudolf, 1891–1970. |
Quine, W. V. (Willard Van Orman) 1908–2000
CLASSIFICATION: LCC B808.5 .P48155 2023 | DDC 146/.4–dc23/eng/20220808
LC record available at https://lccn.loc.gov/2022024945

ISBN 978-1-108-49424-3 Hardback

Cambridge University Press has no responsibility for the persistence or accuracy of URLs for external or third-party internet websites referred to in this publication and does not guarantee that any content on such websites is, or will remain, accurate or appropriate.

For my teachers

Contents

List of Contributors   *page* ix
Acknowledgments   xi

Introduction   1

PART I   CARNAP, QUINE, AND LOGICAL EMPIRICISM

1. Carnap and Quine: First Encounters (1932–1936)   11
   Sander Verhaegh

2. On Quine's Guess about Neurath's Influence on Carnap's *Aufbau*   32
   Thomas Uebel

3. Frameworks, Paradigms, and Conceptual Schemes: Blurring the Boundaries between Realism and Anti-Realism   52
   Sean Morris

PART II   CARNAP, QUINE, AND AMERICAN PRAGMATISM

4. Pragmatism in Carnap and Quine: Affinity or Disparity?   73
   Yemima Ben-Menahem

5. Objectivity Socialized   92
   James Pearson

6. Whose Dogmas of Empiricism?   114
   Lydia Patton

PART III   CARNAP AND QUINE ON LOGIC, LANGUAGE, AND TRANSLATION

7. Reading Quine's Claim that Carnap's Term "Semantical Rule" Is Meaningless   135
   Gary Ebbs

8	What Does Translation Translate? Quine, Carnap, and the Emergence of Indeterminacy *Paul A. Roth*	154
9	Quine and Wittgenstein on the Indeterminacy of Translation *Andrew Lugg*	177
10	Turning Point: Quine's Indeterminacy of Translation at Middle Age *Richard Creath*	194

PART IV CARNAP AND QUINE ON ONTOLOGY AND METAPHYSICS

11	Carnap and Quine on Ontology and Categories *Roberta Ballarin*	215
12	Carnap and Quine on the Status of Ontology: The Role of the Principle of Tolerance *Peter Hylton*	235
13	Carnap, Quine, and Williamson: Metaphysics, Semantics, and Science *Gary Kemp*	253

Bibliography	272
Index	289

Contributors

ROBERTA BALLARIN is Associate Professor of Philosophy at the University of British Columbia.

YEMIMA BEN-MENAHEM is Professor (Emerita) of Philosophy at the Hebrew University of Jerusalem.

RICHARD CREATH is President's Professor of Life Sciences and of Philosophy at Arizona State University.

GARY EBBS is Professor of Philosophy at Indiana University.

PETER HYLTON is Emeritus Distinguished Professor of Philosophy at the University of Illinois at Chicago and Research Professor at Boston University.

GARY KEMP is Senior Lecturer in Philosophy at the University of Glasgow and a member of the Philosophical Faculty of the University of Hradec Králové.

ANDREW LUGG is Emeritus Professor of Philosophy at the University of Ottawa.

SEAN MORRIS is Professor of Philosophy at Metropolitan State University of Denver.

LYDIA PATTON is Professor of Philosophy and Affiliate in the Department of Science, Technology, and Society at Virginia Tech.

JAMES PEARSON is Associate Professor of Philosophy at Bridgewater State University.

PAUL A. ROTH is Distinguished Professor of Philosophy at the University of California, Santa Cruz.

THOMAS UEBEL is Emeritus Professor of Philosophy at the University of Manchester.

SANDER VERHAEGH is Assistant Professor at the Tilburg Center for Logic, Ethics and Philosophy of Science at Tilburg University.

Acknowledgments

I would like to thank Hilary Gaskin at Cambridge University Press for her interest, encouragement, and patience with this project. I would also like to thank the Department of Philosophy and the College of Liberal Arts and Sciences at Metropolitan State University of Denver for their support of this project since its beginnings as a conference in fall 2017. I am also greatly indebted to the contributors. This volume would, of course, not have been possible without them. I am grateful for their willingness to contribute and for their patience over the rather challenging past year or so. I have dedicated it to my teachers and am very happy that some of them are among the contributors. Finally, I thank Haewon and John. There is very little I could do without their support.

Introduction

Rudolf Carnap (1891–1970) and W. V. Quine (1908–2000) are arguably the most important philosophers of the analytic tradition in the twentieth century. A fairly standard account of their respective places in the development of analytic philosophy is that Carnap rose to prominence in the first half of the century, and by 1950, he and the other logical empiricists had become the dominant trend in analytic philosophy, especially in the philosophy of science. Typically, Carnap and the other logical empiricists have been taken as the heirs to Russell and British empiricism more generally. This story continues with Quine coming to prominence in the second half of the century with his refutation of the analytic/synthetic distinction in his 1951 "Two Dogmas of Empiricism." With this paper (along with the 1962 publication of Thomas Kuhn's *The Structure of Scientific Revolutions*), Quine displaced logical empiricism as the dominant trend in analytic philosophy and his naturalism took over, reopening the path to the analytic metaphysics that Carnap and his followers had so railed against. Carnap and Quine then are largely seen as the opposing halves of twentieth-century analytic philosophy.

Recent scholarship has however brought into question this rather simplistic account of Carnap and Quine's philosophical relationship and their roles in the development of analytic philosophy, though it has in no way decreased their importance to the tradition. Much of Quine's work is still viewed as a response to Carnap with much of the tradition that followed the publication of "Two Dogmas" being viewed as a response to Quine. The standard account, however, has been problematized in a number of ways. First, beginning in the 1980s Carnap's previously thought failed philosophical program began to see some important reassessment. In particular, the dominant reading – due largely to Quine himself – of Carnap's *Aufbau* as the culmination of Russell and British empiricism has been shown to be flawed. Instead, the work has been shown to emerge out of Carnap's neo-Kantian background and to have little to do with traditional

Cartesian epistemological concerns with certainty. Similarly, Carnap's appeals to and account of the analytic/synthetic distinction have also been reconsidered. On the standard account, his appeal to the distinction has often been understood – again due in no small part to Quine's reading of it – as providing empiricism with an account of mathematical certainty. Recent scholarship has again shown Carnap standing outside of such traditional epistemological concerns and instead replacing epistemology with what he described instead as the logic of science. His appeal to the analytic/synthetic distinction was all part of his attempt to give a proper methodological and anti-metaphysical account of the sciences.

Quine, due perhaps to a philosophical career that lasted until his death in 2000, has typically been treated as more of a contemporary analytic philosopher than as an historical figure. But more recently, he, too, has been subjected to focused historical attention. In particular, he has emerged perhaps as a much more systematic philosopher – his entire body of philosophical work being understood now as part of a project to work out his naturalism – than previously thought. Also, emerging has been the influence on him by philosophical predecessors such as C. I. Lewis, Bertrand Russell, and, of course, Carnap. What has emerged is that Quine is perhaps better understood in dialogue with these philosophers than with much of the analytic philosophy that followed them and that to a significant degree, he gave rise to.

Despite the historical reevaluation of each of Carnap and Quine, they are still generally viewed as philosophical opponents – again, with their dispute over the analytic/synthetic distinction taking center stage. The essays in this volume – in various ways – question the interpretive strategy of viewing them primarily as opponents. Instead, it is urged here that the many and important differences between Carnap and Quine can be better understood by viewing them as largely sympathetic to each other with their differences emerging as they both strove to achieve a scientific approach to philosophy. For example, the analytic/synthetic distinction then emerges as a question of whether it can be made properly scientific sense of. The approach of this volume to Carnap and Quine is not entirely without precedent. One of the most important early examples of this approach is Richard Creath's 1987 "The Initial Reception of Carnap's Doctrine of Analyticity," where he describes Quine's early engagement with Carnap over the notion of analyticity as a largely sympathetic one. Despite many views to the contrary, Creath shows that Quine's 1936 "Truth by Convention" is not so much an early rejection of the analytic/synthetic distinction as an attempt to work out certain weaknesses that Quine saw in Carnap's account of analyticity. Another more recent example of this approach is Gary Ebbs's 2014 "Quine's Naturalistic Explication of Carnap's Logic of Science." Many commentators have distinguished Quine from Carnap by seeing Quine as attempting to carry out something of the traditional empiricist idea that our best theories of nature are justified by sensory evidence and so, are likely to be true. Instead, Ebbs argues that Quine's epistemology can only be understood by seeing him as incorporating while also transforming Carnap's rejection of this traditional empiricist view. Many other examples also exist, including Ricketts (2004), Hylton (2013), and Verhaegh (2018). While there are many precedents for the approach urged in this volume, this is the first single collection to take it as a basic interpretive strategy for understanding Carnap and Quine, both as individual philosophers and in relation to each other. Previously, this approach could only be found scattered through the literature on these two figures. It is the aim of this volume to bring together early proponents of this approach with some more recent interpreters to demonstrate how looking at Carnap and Quine as figures sympathetic to each other is a profitable interpretive strategy not just for bringing to light the sort of deep philosophical commitments that they shared but also for better understanding the nature of their disagreements.

The volume is divided into four parts.[1] The first, "Carnap, Quine, and Logical Empiricism," considers the varied relations that Carnap and Quine had to logical empiricism. Setting the stage for many of the papers that follow, Sander Verhaegh, in his "Carnap and Quine: First Encounters (1933–1936)," turns to the very beginning of Carnap and Quine's philosophical relationship, examining Quine's visit to Europe during academic year 1932/33, during which he spent five weeks in Prague with Carnap. Verhaegh details what initiated Quine's trip, the events leading up to his arrival in Prague, and finally the momentous philosophical exchange between Quine and Carnap that began there and that would carry on for the rest of Quine's career, even after Carnap's death in 1970. Following up on Verhaegh's contribution, is Thomas Uebel's "On Quine's Guess about Neurath's Influence on Carnap's *Aufbau*." Here again we have Quine in dialogue with Carnap but this time by way of fellow logical empiricist, Otto Neurath. Uebel focuses on some speculations of Quine's about how Neurath's criticisms of Carnap's *Aufbau* gave rise to Carnap's physicalism. Particularly relevant to the aims of this volume is Uebel's account of how

---

[1] In the summary of the volume that follows, I have in places drawn heavily from contributors' submitted abstracts. For this, I thank them for their assistance with this Introduction.

all of Neurath, Carnap, and Quine ultimately converge upon physicalism, though by different routes. Part I of the volume concludes with my contribution, "Frameworks, Paradigms, and Conceptual Schemes: Blurring the Boundaries between Realism and Anti-Realism." Here, I argue that despite Quine and Kuhn's reputation for bringing to a close the era of Carnap and logical empiricism as dominating philosophy of science, all of Carnap, Quine, and Kuhn share in rejecting traditional realist and anti-realist analyses of ontology. They all reject a version of what Putnam called metaphysical realism and then also resist making the move to an equally extra-scientific anti-realist position. For all of them, science itself is to be the final arbiter of what there is.

Part II of the volume turns to consider Carnap and Quine in relation to American pragmatism, as a possible common philosophical influence on each of them. Yemima Ben-Menahem begins the discussion with her "Pragmatism in Carnap and Quine: Affinity or Disparity?" noting that Quine's remark from the end of "Two Dogmas of Empiricism" concerning the pragmatism of himself and Carnap invites drawing a connection to American pragmatism. She carefully distinguishes, however, Quine's use of "pragmatism," which has a rather general sense, from the more specific features of American pragmatism. Only with this distinction carefully in place can we then get an accurate sense of the influence of American pragmatism on each of Carnap and Quine. Perhaps surprisingly, she concludes that the pragmatist influence, particularly that of William James, may have been stronger on Quine than it was on Carnap despite Quine's own disavowals of it. We next turn to James Pearson's "Objectivity Socialized." Here, Pearson examines a challenge from Charles Morris to Carnap's philosophy and a related challenge from Donald Davidson to Quine's that they both neglect the social nature of inquiry. Ultimately, Pearson argues that those of us who wish to pursue Carnap and Quine's scientific vision of philosophy must recognize the ineliminable role that other inquirers play in our own investigations. Lydia Patton concludes Part II with her "Whose Dogmas of Empiricism?" Building on recent work that emphasizes certain broad areas of agreement between Carnap and Quine, Patton questions whether Carnap was really Quine's main target in his "Two Dogmas of Empiricism." Examining the influence of pragmatism on both Carnap and Quine, she concludes that Quine's opposition to the "conceptual pragmatism" of Clarence Irving Lewis was a deeper motivation for his positions in "Two Dogmas" than his opposition to Carnap's linguistic defense of analyticity.

Part III looks at Carnap and Quine's views on logic and language with a particular emphasis on the notion of translation. Gary Ebbs opens this section with his "Reading Quine's Claim that Carnap's Term 'Semantical Rule' Is Meaningless." According to Ebbs, many informed readers of Carnap (and Quine) have taken Quine's objections to Carnap's account of analyticity in terms of semantical rules to have failed. Ebbs counters this, arguing that Quine actually saw himself as applying Carnap's own philosophical standards more strictly than Carnap himself did. Quine was, as he later reported, "just being more carnapian than Carnap." Ebbs's careful analysis of Section 4 of "Two Dogmas of Empiricism" shows Carnap conflating two senses of "semantical rule." Although the first is clear, Quine sees it as being of no use in defining analyticity. The second, though integral to Carnap's method of defining analyticity, Quine shows to be left unexplained by Carnap's definitions. Paul Roth then turns the discussion toward the indeterminacy of translation with his "What Does Translation Translate? Quine, Carnap, and the Emergence of Indeterminacy." Roth observes that both Carnap and Quine see an element of practical choice in our scientific theorizing but that they diverge on its significance, particularly with regard to a theory of meaning. From Carnap's standpoint, linguistic frameworks are practically adopted without any prior constraints and then provide for a theory of meaning. In contrast, Quine sees a theory of meaning presupposing a more general assumption that all meaningful elements stand in a systematic relation before translation begins. Without this assumption, there is no work for a theory of meaning to do. In this sense, the dogmas of empiricism can only do explanatory work if the meaningful elements are already systematically linked in a way that translation might recapture. Next, in his "Quine and Wittgenstein on the Indeterminacy of Translation," Andrew Lugg examines the similarity of Quine's indeterminacy thesis with views of Wittgenstein's, arguing that Quine and Wittgenstein are, for all their differences, reasonably regarded as battling a commonly held philosophical conception of the determinateness of translation. In conclusion, Lugg brings us back to Carnap – perhaps Wittgenstein's most important successor and Quine's most important predecessor – and argues that he largely agreed with Quine and Wittgenstein, his reservations about many of their views notwithstanding. Richard Creath concludes this part of the volume with his "Turning Point: Quine's Indeterminacy of Translation at Middle Age." He argues that Quine's indeterminacy of translation as presented in his *Word and Object* (1960) represents a turning point in his thinking. While Quine may have started out as a disciple of Carnap's, in the 1940s and '50s the most salient feature of Quine's work is a deep asymmetry between intensional and extensional concepts. Creath argues that the arguments for the indeterminacy of translation undermine this asymmetry and initiate changes to the role of ontology and reference, to the status of simplicity, to Quine's understanding of analyticity and synonymy, and to the character and centrality of his epistemology, ultimately including even a return to a two-tier epistemology. While the changes do not amount to a wholesale rejection of earlier views, in the aggregate, they were significant and brought Quine's position back much closer to Carnap's.

Part IV concludes the volume by examining Carnap and Quine's views on ontology and metaphysics. Roberta Ballarin begins this section with her "Carnap and Quine on Ontology and Categories," joining recent debates over Quine's understanding of Carnap's "Empiricism, Semantics, and Ontology." While a number of commentators have argued that Quine's account of Carnap's paper in terms of the category/subclass distinction is simply a misunderstanding of Carnap, Ballarin argues that it is not. Instead, she argues that Quine was correct to construe Carnap's external questions of existence as all being category questions. She then, however, dissents from a second claim of Quine's – that answers to internal category questions of existence are trivial and analytic. Here, she dissents from a view of Ebbs, who has recently argued that Quine was right on both points. Ballarin shows that epistemic considerations that support Quine's first point, undermine his second point. Next, Peter Hylton takes up Carnap and Quine's views on ontology in his "Carnap and Quine on the Status of Ontology: The Role of the Principle of Tolerance." While both Carnap and Quine see their disagreement over the status of ontology as a legitimate philosophical undertaking as ultimately rooted in their disagreement over the analytic/synthetic distinction, Hylton argues that this cannot be so since Quine comes to accept a notion of analyticity without changing his views on ontology. Instead, Hylton argues that the more fundamental point underlying the disagreement about the status of ontology is Carnap's advocacy of the Principle of Tolerance, which Quine never comes to accept. Finally, Gary Kemp concludes the volume with his "Carnap, Quine, and Williamson: Metaphysics, Semantics, and Science." While Quine is often taken to have broken the Viennese straitjacket of Logical Positivism, which rejected metaphysics, as an a priori but non-analytic, substantive discipline, allowing speculative metaphysics to be reborn, Kemp thinks this is incorrect.[2] As he explains, for all their much-discussed disagreements over analyticity and ontology, Quine shared

---

[2] The rather picturesque phrasing of the "the Viennese straitjacket" is Gary Kemp's.

Carnap's more fundamental commitment to "scientific philosophy": to the idea that legitimate philosophy is the work of handmaidens, site managers or accountants of science. Their primary role is to act to clarify, precisify and make explicit the methods and deliverances of science. Kemp then brings Carnap and Quine to bear on more recent analytic trends towards metaphysics by specifically contrasting Carnap and Quine's scientific philosophy with recent work by Timothy Williamson. Kemp stresses Carnap and Quine's considerable distance from Williamson; and that from Quine's point of view as well as from Carnap's, this recent ascendance of metaphysics will seem a departure from science without sufficient justification.

PART I

Carnap, Quine, and Logical Empiricism

CHAPTER I

Carnap and Quine
First Encounters (1932–1936)

Sander Verhaegh

1.1 Introduction

Carnap and Quine first met in the 1932–33 academic year, when the latter, fresh out of graduate school, visited the key centers of mathematical logic in Europe. In the months that Carnap was finishing his *Logische Syntax der Sprache*, Quine spent five weeks in Prague, where they discussed the manuscript "as it issued from Ina Carnap's typewriter" (Quine 1986a, 12). The philosophical friendship that emerged in these weeks would have a tremendous impact on the course of analytic philosophy. Not only did the meetings effectively turn Quine into a Carnapian "disciple" (Quine 1970a, 41), they also paved the way for their seminal debates about meaning, language, and ontology – the very discussions that would change the course of analytic philosophy in the decades after the Second World War. Yet surprisingly little is known about these first meetings. Although Quine has often acknowledged the impact of his Prague visit, there appears to be little information about these first encounters, except for the fact that the Quines "were overwhelmed by the kindness of the Carnaps" and that it was Quine's "most notable experience of being intellectually fired by a living teacher" (Quine 1985, 97–98). Neither their correspondence (Quine and Carnap 1990, 108–120) nor their autobiographies (Carnap 1963a; Quine 1985, 97–98; 1986a, 12–13) offer a detailed account of these meetings.

In this essay, I shed new light on Carnap's and Quine's first encounters by examining a set of previously unexplored material from their personal and academic archives.[1] Why did Quine decide to visit Carnap? What did

---

I would like to thank Peter Hylton, Nathan Kirkwood, and Sean Morris, as well as audiences in Chicago and Glasgow for their valuable comments and suggestions. This research is supported by The Netherlands Organization for Scientific Research (grant 275-20-064).

[1] In addition to material from the Rudolf Carnap Papers at Pittsburgh's Archives of Scientific Philosophy (hereafter, RCP) and the W. V. Quine Papers at Harvard's Houghton Library (hereafter, WVQP), the present paper is based on a study of fifty-seven boxes of new material (private

they discuss? And in what ways did the meetings affect Quine's philosophical development? In what follows, I address these questions by means of a detailed reconstruction of Quine's year in Europe based on a range of letters, notes, and reports from the early 1930s.[2]

1.2 Cambridge

Quine visited Europe between September 1932 and June 1933, a trip that was funded by a Frederick Sheldon Traveling Fellowship. At the time, a year in Europe was by no means unusual for Harvard's best and brightest. Already in his first year in graduate school, Quine sketched the Europe route to a professorship in a letter to his parents:

> I feel as though [I] have a good chance spending the year after next in Europe ... The usual thing for the favored few here seems to be: get Ph.D., then be sent to Europe ... then come back and be an instructor here at Harvard for a year, and then pick your place! (May 27, 1931, DBQ21)

Quine was well aware that he was one of Harvard's "favored few." Some of the most prominent philosophers residing in Emerson Hall – A. N. Whitehead, C. I. Lewis, and Henry Sheffer – were clearly fond of the ambitious logician, who was trying to complete both his M.A. *and* his Ph.D. in two years. In several letters written during his first year of graduate school, Quine speaks about his excellent "stand-in with Sheffer, Whitehead and Lewis" (May 21, 1931, DBQ21). Indeed, Whitehead told Quine that he "was the first pupil he had ever had whom he believed to understand exactly what they [Russell and Whitehead] had been up against in the *Principia*" (March 16, 1931, WVQP, Item 1215).

Quine did not only seek a Sheldon Fellowship in order to boost his chances on the job market. There were also good *philosophical* reasons to visit Europe. Quine had come to Harvard in September 1930, after graduating from Oberlin College with a major in mathematics and honors in mathematical philosophy. At the time, he believed that Harvard would be the best place for an aspiring logician; he had extensively studied Whitehead's *Introduction to Mathematics* and read about Sheffer's stroke function in Russell's *Introduction to Mathematical Philosophy* (Quine 1986a, 7–8).

---

manuscripts, date books, and nonacademic correspondence) made available by Quine's son and literary executor Douglas B. Quine. I will refer to documents from this additional collection by listing box numbers preceded by the marker DBQ. Transcriptions are mine unless indicated otherwise.

[2] Due to limitations of space, this paper mostly discusses the first encounters from Quine's perspective. For a reconstruction from Carnap's perspective, see Verhaegh (2020b).

During his first year at Harvard, however, Quine quickly discovered that Whitehead was not teaching logic and that Sheffer was mainly talking about "peripheral papers." In his autobiography, Quine remembers:

> American philosophers associated Harvard with logic because of Whitehead, Sheffer, Lewis, and the shades of Peirce and Royce. Really the action was in Europe. In 1930 and 1931, Gödel's first papers and Herbrand's were just appearing, but there were already other notables to reckon with: Ackermann, Bernays, Löwenheim, Skolem, Tarski, von Neumann. Their work had reached few Americans. (Quine 1985, 83).

Quine's last remark appears to be somewhat of an exaggeration. His reading list for Sheffer's course on "relational logic" in the 1930–31 academic year (WVQP, item 3237) shows that Sheffer was quite up to date when it came to developments in logic on the continent. Quine's notes of Sheffer's first lecture show that they were not only discussing Wittgenstein's *Tractatus*, they also reveal that the students were reading Carnap's *Abriss der Logistik*, a book that had been published only a year before.

Still, it seems correct that Harvard philosophers gradually started to realize that the "action was in Europe" when Quine entered graduate school. For this was exactly the period in which Herbert Feigl, one of the core members of the Vienna Circle, visited Harvard on a Rockefeller Fellowship and started to spread the *Wissenschaftliche Weltauffassung* in the United States.[3] Feigl's correspondence from the early 1930s shows that he introduced the Viennese views to Sheffer, Whitehead, and Lewis in Harvard colloquia and that he played an important role in advertising logical positivism at meetings of the American Philosophical Association. Indeed, by the end of the 1930–31 academic year, Lewis was already describing logical positivism as "what we in America are sure to regard as the most promising of present movements in Continental philosophy."[4]

Considering Feigl's active promotion of the views of the *Wiener Kreis* at Harvard, it is not surprising that Quine, in writing an application for a Sheldon Fellowship, decided to spend most of his time in Vienna. Already in the above-mentioned letter to his parents, Quine mentions that there is "an active school of logicians at Vienna" and that this would "be the place where I'd do most of my studying" (May 27, 1931, DBQ21). In fact, a 1931 letter reveals that Feigl also played an important role in convincing

---

[3] See, for example, Blumberg and Feigl's manifesto "Logical Positivism: A New Movement in European Philosophy," published in May 1931.
[4] April 14, 1931, HF 03-53-01, Herbert Feigl Papers, University of Minnesota Archives. See Verhaegh (2020a) for a reconstruction of Feigl's year at Harvard.

Quine that he should visit Prague to talk to Carnap. According to Feigl, meeting Carnap would be indispensable for an aspiring mathematical philosopher:

> Our best logician, *Carnap* (his highly important contributions to mathematical logic ... ha[ve] not been published yet) has moved to Prague ... He knows and lectures a lot, too, about Foundations of Math. – I would advise you to see him at any rate. (Feigl to Quine, December 1931, WVQP, Item 345, original emphasis)

In fact, Feigl ends his letter with a list of the five cities Quine should try to visit if his fellowship allows it, including Prague on the top of his list:

(1) *Prague* (with Carnap only, nothing else worthwhile)
(2) *Berlin* (J. v. Neumann, the most brilliant Hilbertian, and try Reichenbach in Space and Time, Probability)
(3) *Warsaw* (the Polish logicians; Lukasiewicz, Lesniewski, Tarski, etc.)
(4) *Vienna* (Schlick, and some younger men like Gödel and Waismann You know, Gödel has *proven* the incompleteness of *any* postulate system for arithmetics. Waismann is the best interpret of Wittgenstein's cryptical philosophy.)
(5) *Cambridge* (England), where you can hear the great prophet himself. (December 1931, WVQP, Item 345, original emphases)

Quine seems to have taken Feigl's advice to heart. He ended up going to Vienna (September 15, 1932–February 28, 1933), Prague (March 1–April 6, 1933), and Warsaw (May 6–June 7, 1933).[5] In his autobiography, Quine mentions that he made inquiries about going to Berlin but that he removed it from his list because it "had nothing to offer in logic" (1985, 94). Quine's correspondence with his parents shows that he also wrote a note to Wittgenstein in order to get an "audience with the prophet" (September 20, 1932, DBQ21). Unsurprisingly, however, Quine never received an answer.[6]

1.3 Quine's Early Development

Initially, Quine and his wife planned to leave the United States in June 1932 and to spend the summer in Europe before the start of the Viennese academic year. Quine had to postpone his steamship reservations, however, when he received the happy news that Harvard's

---

[5] See Quine's date books for 1932 and 1933 (DBQ45) and his "General Report of my Work as a Sheldon Traveling Fellow" (January 8, 1934, WVQP, Item 3254).
[6] See Quine (1985, 87–88).

department of philosophy had decided to subsidize the publication of his dissertation. If he wanted to get his book published during his year abroad, he had to rework the manuscript and prepare it for printing before he left for Europe.

Harvard's decision to subsidize the publication of the thesis probably did not come as a surprise. In logic, Quine knew, the most prominent players held his work in high regard. Whitehead, who had supervised the thesis, was particularly impressed. In a 1933 recommendation letter, he described Quine as one of the most talented logicians he had ever worked with:

> In the course of 45 years of experience, the only two men who at his age – 25 yrs – submitted comparable work were Maynard Keynes and Bertrand Russell. And his superior common sense gives him the advantage over the latter. (February 24, 1933, WVQP, Item 1215)

Philosophically, however, Quine's position was still largely in development. His papers and notebooks from the early 1930s reveal that he defended a somewhat unusual combination of behaviorist and phenomenalist views in epistemology.[7] On the one hand, Quine was convinced that behaviorist analyses of mind and language provide the tools to solve a great many problems in the theory of knowledge. Already in his student years, Quine defended an epistemology in which our knowledge of the external world is viewed a web of sentences,[8] some more deeply entrenched than others, that we have come to accept through the processes of psychological conditioning. In addition, he defended a holistic perspective on theory revision, arguing that the inquirer "has a certain latitude as to where he may make his readjustments in the event of an experience recalcitrant to his system" (March 10, 1931, WVQP, Item 3236). Prima facie, these views seem to be remarkably similar to the epistemology Quine first outlined in "Two Dogmas of Empiricism." In the early 1930s, however, Quine's holism did not extend to logical and mathematical knowledge. Still, he seems to have accepted that our knowledge of logic, too, should somehow be accounted for in behaviorist terms. In a note titled "The Validity of Deduction," Quine argued that logical concepts "bear to external reality merely the relation of psychological response to stimulus" (April 11, 1930, WVQP, Item 3224) and he combined this with the Lewisian view that the so-called "eternal validity of logic ... is nothing more

---

[7] The following two paragraphs build on Verhaegh (2018) and (2019a).
[8] Quine used the web metaphor as early as 1927. See "On the Organization of Knowledge" (March 19, 1927, WVQP, Item 3225).

than ... the property of a definition to remain immutable" unless "altered by convention" (April 11, 1930, WVQP, Item 3224).⁹

On the other hand, Quine also accepted a variant of phenomenalism. Although Quine, again like Lewis, criticized naïve sense data theories,[10] he did maintain that every theory, including the theories of the behaviorist, ultimately requires an epistemological, phenomenalist basis. In a paper for a course by Whitehead, Quine wrote:

> It may well suit the purposes of the neurologist or psychologist to ... [explain their theories in term of] conditioned reflexes and general habit responses; but it must be remembered that such treatment ... depends upon the prior adoption of a whole system of concepts and hypotheses. Philosophy, if it would inquire into the nature of all such conceptual systems and hypotheses, must certainly endeavor to remain aloof from the initial adoption of any one such system ... let the psychologically prime be what the psychologist finds most efficacious; for philosophy, no one item is initially certified as of more fundamental or ultimate character than any other. I am driven, therefore, to identifying the "bare datum" with that which Professor C. I. Lewis calls "the given." (March 10, 1930, WVQP, Item 3225)

This tension between (1) a behavioristic epistemology and (2) a phenomenalist perspective according to which behaviorism is just one "system of concepts" among many is a constant in Quine's early career.[11] Quine seemed to be caught between two competing perspectives, neither of which was fully satisfactory. Behaviorist analyses of knowledge seemed to ignore valid questions about the epistemic status of the behaviorist theory itself, whereas phenomenalist perspectives appeared to ignore the fact our theory of the world is just a system of concepts and sentences we have come to accept through the processes of psychological conditioning.

1.4 Vienna

The Quines boarded *The President Roosevelt* in August 1932 and spent the first weeks of their year in Europe in and around France. Quine had

---

⁹ I describe Quine's conventionalist leanings as 'Lewisian' because it appears to be inspired the view that we 'create' necessary truths by making classifications, a theory Lewis defends in *Mind and the World-Order* (1929). Frost-Arnold (2011, 300n15) also suggests that Quine's identification of the a priori with claims that can be held true 'come what may' was influenced by C. I. Lewis. The main difference between Quine and Lewis is that the former formulated this theory in behavioristic terms.

[10] See Quine (March 10, 1930, WVQP, Item 3225): "No analysis of a given experience can yield any other experience which is, in any full sense, the 'bare datum' of the form of experience; any such analysis is, rather, merely a further interpretation."

[11] See Verhaegh (2017, 2018, 2019a, 2019b).

frantically worked on the manuscript for his book until mid-August, completing it just in time to mail it to Lewis a few hours before they took a bus to the New York harbor (Quine 1985, 87). On September 11, they arrived in Vienna, a city they quickly deemed the most beautiful place they had visited in their brief but extensive travel careers. In a letter to his parents, Quine noted that Vienna "surpasses Paris or any other big city," and that they were especially enjoying the beauty of the streets as well as the "many public buildings, parcs, and palaces" (September 14, 1932, DBQ21).

Despite the beautiful city and the happy prospect about the publication of his book, Quine's first months in Vienna were somewhat of a disappointment. Not only did he discover that the university would not be open for seven weeks – a period the Quines used for a short trip to the Balkans; he also had to conclude that there were no courses in mathematical logic, the prime reason for his trip to Europe. In a report of his work as a Sheldon Travelling Fellow, Quine writes:

> I was disappointed ... at the lack of activity in mathematical logic. After much investigation I was informed that no lectures were being given on the subject. Extended inquiries among deans and registrars, as to what might be found in the way of seminars or discussion groups on the subject, all proved futile. (January 8, 1934, WVQP, Item 3254)

In fact, even the philosophy lectures were difficult to attend due to the somewhat chaotic administration at the University of Vienna. Quine was auditing, among others, Schlick's lectures on philosophy but complained that the "professors fail to appear about half the time," noting that the "same sort of frustration attends the use of the library, the quest of information, and every other activity" (November 25, 1932, DBQ21).[12]

In response to these setbacks, the administrative chaos, and the lack of activity in mathematical logic, Quine pondered leaving the Austrian capital and going to Prague or Warsaw straight away as he became increasingly disillusioned by the chaos at the University of Vienna:

> I [have] become impatient with the passive resistance of Vienna and the difficulty of getting anything accomplished ... I am convinced now that there is nothing to hope for here ... Vienna is a keen town, but, if it hadn't been for my much more extreme experiences in the Balkans, I should be inclined to believe that the Austrians were the world's most helpless people. System is unknown. (November 25, 1932, DBQ21)

---

[12] An additional source of disappointment was the news that the publication of his book was severely delayed. See Quine's correspondence with Lewis about the publication process (WVQP, item 1464). In the end, Quine's book was published only in September 1934.

By the end of November, however, Quine finally had a chance to have a meeting with Schlick. And although Quine was "certain that no advice ... could warrant" his "staying in Vienna" (ibid.), a week later he reported that his meeting "changed everything." The German professor had invited him to come to the discussions of the *Wiener Kreis*, informed him that their next meeting was going to be held the very next day, and asked him to give a talk in January:

> The talk with Schlick changed everything ... I had been sure beforehand that he could tell me of nothing encouraging in Vienna [but] he told of this circle of his ... and invited me to come regularly. The next meeting was the very next day ... I went and found that the group numbered about fifteen, practically all middle-aged or elderly men and all apparently people who have already produced something. I had already, in America, heard of several of the names – Schlick, Waismann, Gödel, and the famous mathematician Karl Menger. ... Thus, all in all, there is interest in logic here after all. (December 5, 1932, DBQ21).

On top of that, Quine learned that Carnap would come to Vienna and that the latter was planning on discussing "the last chapter of his next book" with the Circle in a few weeks' time (ibid.). Quine's luck was clearly changing; not only did he finally have "access to the Inner Circle" (January 8, 1934, WVQP, Item 3254), he would also be meeting Carnap, the person who Feigl had described as the Circle's greatest logician.

1.5 *Aufbau*

Quine was not only excited to meet Carnap to learn about his contributions to logic. He was also curious to talk to the German professor because he had been reading the latter's *Der Logische Aufbau der Welt* during his first months in Europe. In August 1932, a few days before he left the United States, Quine had received a thirteen-page letter about the *Aufbau* from John Cooley, a former fellow graduate student, who had read the book and urged Quine to do the same. According to Cooley, Carnap had written a "very ingenious" book, attempting to "use the methods of symbolic logic to work out a strictly positivistic philosophy, more or less on the lines which Russell indicated" in *Our Knowledge of the External World* (Cooley to Quine, August 6, 1932, WVQP, Item 260). Quine, who appears not to have been familiar with Carnap's work beyond the above-mentioned *Abriss der Logistik* (see Section 1.2) and had read Russell's programmatic epistemology as a sophomore must have been intrigued by Cooley's letter.[13] For when he arrived in Vienna, he immediately borrowed a copy from the local library and studied the book during his first weeks in Europe.

In the *Aufbau*, Quine discovered, Carnap attempted to develop a "constructional system of concepts" in which all concepts of the empirical sciences are derived or constructed "from certain fundamental concepts," such that "a genealogy of concepts results in which each one has its definite place" (Carnap 1928a, §1). Just as a system of arithmetical concepts can be created by constructing these concepts, step by step, "from the fundamental concepts of natural number and immediate successor" (§2), Carnap argued, so we can construct all concepts of the empirical sciences "from a few fundamental concepts," most notably the concept of an "elementary experience" (*elementarerlebnisse*), an individual's totality of experiences at a given moment in time involving all sense modalities – or as, Quine summarized it in his reply to Cooley, the "uncontrollable given." Using only the tools of logic, set theory and this "sense-datum language in the narrowest conceivable sense," Quine would later write, Carnap managed to define "a wide array of important additional sensory concepts which … one would not have dreamed were definable on so a slender basis" (1951e, 39).

Present-day Carnap scholars almost unanimously reject this phenomenalist (Russellian) reading of the *Aufbau* and argue that Carnap wanted to *overcome* the subjectivity of sense experience rather than to "account for the external world as a logical construct of sense data" (Quine 1969a, 74).[14] From a historiographical perspective, therefore, it is interesting to note that Quine, likely influenced by Cooley's summary of the project as well as his own phenomenalist leanings (see Section 1.3), interpreted the *Aufbau* in Russellian terms from his very first reading in 1932. In his response to Cooley, for example, Quine writes:

> It seems that Carnap has paved the way for carrying out in detail that to which Russell has merely pointed in his doctrine of 'logical constructions' … The *Aufbau* stands to [the] philosophical doctrines of Russell … as *Principia* stands to the antecedent purely philosophical suggestion that mathematics is a form of logic. (Quine to Cooley, April 4, 1933, WVQP, Item 26)

What Quine did *not* know when he wrote this letter, however, was that Carnap had radically changed his perspective in the fall semester

---

[13] See Quine (1985, 58).
[14] See, for example, Richardson (1998), who argues that the *Aufbau* should be read as a neo-Kantian (rather than as an empiricist) project – that Carnap's notion of "logical form" should be interpreted as a notion of form in the Kantian sense.

of 1932.[15] For in the very weeks that Quine had first started reading the *Aufbau*, Carnap had been writing "Über Protokollsätze," the paper in which he rejected what he by then called the "residue of ... absolutism" in the views of the Vienna Circle:

> In all theories of knowledge up until now there has remained a certain absolutism: in the realistic ones an absolutism of the object, in the idealistic ones (including phenomenology) an absolutism of the 'given' ... There is also a residue of this idealistic absolutism in ... our circle ... it takes the refined form of an absolutism of the ur-sentence ('elementary sentence', 'atomic sentence') ... It seems to me that absolutism can be eliminated. (Carnap 1932c, 469)

Carnap's change of heart had important consequences for his philosophy. He started to view ur-sentences (protocol sentences) as relative, arguing that science does not rest upon solid bedrock (the given) but should be viewed as a building erected on piles driven down into a swamp,[16] and he started to follow Neurath in defending the view that elementary sentences are revisable. If our protocols are not absolute, Carnap maintained, we always have the option to revoke them when they conflict with some of our best-established hypotheses. Most importantly, he changed his metaphilosophical perspective on the question whether or not we ought to start with phenomenalist protocol sentences in the first place. In "Über Protokollsätze," Carnap for the first time argues that this is not a question of a fact but a linguistic decision:

> this is a question, not of two mutually inconsistent views, but rather of *two different methods for structuring the language of science both of which are possible and legitimate* ... possible answers ... are to be understood as suggestions for postulates; the task consists in investigating the consequences of these various possible postulations and in testing their practical utility. (Carnap 1932c, 457–458)

In arguing that the question of what protocol language to adopt is a question of linguistic decision, Carnap was paving the way for his principle of tolerance, a principle which, as we shall see in Section 1.8, would come to be central to his philosophy from 1933 onwards.

1.6 Vienna Circle

Quine could have learned about the changes to Carnap's epistemology in December 1932, when the two had scheduled a meeting to arrange the

---

[15] This paragraph is based on Verhaegh (2020b).
[16] The metaphor is from Popper, who had convinced Carnap about his view in 1932. See Carus (2007a, 253).

details of Quine's Prague visit and Carnap had planned to discuss the last chapter of his book manuscript with the Vienna Circle.[17] Unfortunately, Quine had to wait a few more months before he would hear about Carnap's new approach to protocol sentences. For Carnap fell ill on the day before the meeting of the *Wiener Kreis* and spent most of December in a Viennese hospital. Quine briefly visited Carnap in the infirmary to wish him a speedy recovery and to arrange the details of his Prague visit[18] but their first *philosophical* encounter had to be postponed until the spring semester, when Quine, the two decided, was going to spend a month in Prague to discuss logic and philosophy.

Perhaps as a result of his first meeting with Carnap, Quine came to view his decision to stay in Austria as a mistake. For although his extended stay in Vienna gave him a chance to attend the weekly meetings of the *Wiener Kreis* as well as to give a talk to the group himself, he does not seem to have hit it off with the members of Schlick's Circle. In reports about his experiences in Austria, Quine even complains about "the dearth of … opportunities for discussion" (January 8, 1934, WVQP, Item 3254) and that the "foreign visitor tends on the whole to be ignored by the Viennese Faculty" (October 20, 1933, WVQP, item 2915).[19] In fact, even the meetings of the Circle itself turned out to be somewhat of a disappointment. Despite Quine's initial enthusiasm about the group's active interest in mathematical logic, he quickly deemed that the meetings were mostly concerned with philosophy:

> The meetings [of the Vienna Circle] were only of moderate interest: Each was occupied by a paper followed by discussion. The meetings proved to be rather philosophical than logical … It was obviously a mistake to have stayed so long in Vienna … It was not until Prague … that I realized how great advantages a traveling fellow might enjoy.[20] (January 8, 1934, WVQP, Item 3254)

Rather than discussing logic with Gödel, Menger, and Schlick, Quine mostly spent his last months in Austria developing a "neater and simpler form of notation" for his forthcoming book. Meanwhile, Quine was looking forward to his visit to Carnap and decided to postpone a trip to Italy he and Naomi had planned in order to get to Prague before the Easter break (February 11, 1933, DBQ21).

---

[17] See Carnap's letter to Quine from December 5, 1932 (Quine and Carnap 1990, 108).
[18] See Quine's letter to his parents, February 11, 1933, DBQ21.
[19] In his 1934 report about his trip to Europe, Quine writes that he only "contrived a few minutes of discussion with [Gödel] after the close of some of the meetings" (January 8, 1934, WVQP, Item 3254). The one person with whom Quine seems to have been talking regularly was another visitor: Alfred Ayer. Quine's letters and datebooks show that the two regularly got together in the first months of 1933.
[20] Quine offers a similar complaint in a letter to Sheffer (February 16, 1933, WVQP, item 981).

1.7 Prague

The Quines arrived in Prague on the first day of March, a month that would prove disastrous for European democracy. For Quine visited Prague in the very month that the German Reichstag passed the *Ermächtigungsgesetz*, the amendment that effectively transformed Adolf Hitler's government into a full-blown dictatorship and is widely viewed as the end of the Weimar Republic. In Austria, too, parliamentary rule was abolished in March 1933. Engelbert Dollfuss, the Austrian Chancellor of the Christian Social Party, took advantage of a procedural hiccup and declared that the Austrian parliament had abolished itself, preventing members of the opposition from entering the chamber.

The rapidly increasing political tensions in Europe had not gone unnoticed to the Quines, who had been living in a radically divided Vienna for more than five months. Indeed, the political situation in Europe was the main topic of a speech he gave at a Harvard philosophy faculty reception in October 1933, a few months after he returned to Cambridge. In his speech, Quine mostly recounts the grim atmosphere in Vienna:

> We witnessed many Nazi parades and demonstrations in Vienna and in small cities in the neighborhood ... Swastikas and anti-Semitic mottoes were painted on walls throughout the city, and from time to time the sidewalks would be strewn with paper swastikas and bits of papers printed with the injunction not to buy from Jews. (October 20, 1933, WVQP, Item 2915)

In Germany, briefly visited by the Quines on their way back to the United States in June, the situation had been even worse. In his speech, Quine recounts that there was "an abundance of Nazi uniforms in the trains and in the streets" and that "Hitler's photograph" was hanging "in practically every show window and on the wall of every café" (ibid.). The Quines were particularly shocked when they learned that even some of the people they frequently interacted with had fallen for the Nazi rhetoric. In his speech, Quine tells an anecdote about the wife of Jan Lukasiewicz, who had expressed her sympathy for Hitler on a few occasions.[21]

Despite the increasingly hostile political situation in Europe, Quine's month in Prague would *intellectually* be the most important period in his early philosophical development. In a report about his Sheldon Fellowship, Quine recounts that "Prague was the antithesis of Vienna," arguing that

---

[21] A few years later, Quine would refuse to help Lukasiewicz to get a post at Harvard because of his suspicions of "Lukasiewicz's relations with the Nazis in the days just preceding the destruction of Poland" (Quine to Kline, October 21, 1945, WVQP, item 588). See also Quine to Stone (December 24, 1945, WVQP, item 659).

his meetings with "Carnap ... alone would have been academic justification" for the entire year (January 8, 1934, WVQP, Item 3254). Not only did Carnap prove to be "a master of classroom technique" in his logic courses (October 20, 1933, WVQP, Item 2915), he was also very interested in Quine's work, inviting him to present his work to the students. On top of that, the Carnaps were incredibly welcoming and friendly, helping Quine and Naomi to find accommodation and inviting them over regularly for drinks and dinner:

> We've been overwhelmed by the solicitude of the Carnaps ... When I talked with him after his first class, Thursday, he invited us out to his place Saturday for tea. Next day ... Mrs. Carnap met us ... and tramped through the streets with us for over three hours ... helping us find a room. Saturday afternoon, when we were out at their house ... Mrs. Carnap had made all manner of fancy and very time consuming pastries for the occasion. When we left they both put on their boots and conducted us through the dark down a steep, muddy field to the bus-line, the four of us slipping down the soft hillside as if we were on skis on a snowy mountain. Such is the great Carnap. (March 7, 1933, DBQ21)

For our present purposes, however, it is especially important that Carnap and Quine also regularly got to together to discuss logic and philosophy. For, in addition to Carnap's five hours of lectures each week and Quine's private study of the book manuscript that Carnap had lent him, Quine recounts that the two had eight meetings of "three to six hours" in the five weeks that the Quines spent in Prague (January 8, 1934, WVQP, Item 3254).[22] Already in his first week in Czechoslovakia, Quine decided that "Carnap's stuff" was "so fruitful" that he could best spend his "time in Prague ... completely mastering Carnap's ideas" (March 7, 1933, DBQ21).

1.8 Syntax

Carnap, we have seen, had extensively revised his attitude to epistemology in the fall semester of 1932. The manuscript that Carnap was finishing in March 1933, however, was related to a different but connected series of developments – to a set of philosophical breakthroughs that fundamentally changed Carnap's metaphilosophy. Before we turn to Quine's response to

---

[22] Besides Carnap's lectures they appear to have had meetings on March 4, 6, 11, 18, 21, 22, 31, and April 4. See Carnap's diary (RCP, 025-75-11) and Quine's letters (DBQ21) and date books (DBQ45).

Carnap's manuscript, therefore, it is useful to discuss the most important advances of what has been called Carnap's *Syntax* period.[23]

The *Syntax* project started in January 1931. In his *Intellectual Autobiography*, Carnap recounts a sleepless, feverish night, during which "the whole theory of language structure" came to him "like a vision" (1963, 53). Up to this night, Carnap had been severely struggling with Wittgenstein's *Tractarian* restriction that we cannot meaningfully talk about the logical form of language. Since (1) the picture theory of meaning implies that a proposition and that what is pictured must share a logical form and (2) that the theory of types prohibits propositions that are speaking about themselves, we have to conclude that the logical form of a proposition itself cannot be represented by a proposition.[24] What Carnap realized, however, is that we do not need to presuppose that propositions about logical form are *empirically* meaningful. If we stick to talking about "the forms of the expressions of a language, the form of an expression being characterized by the specification of the signs occurring in it and of the order in which the signs occur" and if one can show that central concepts of metalogic (e.g., logical consequence, derivability) are purely syntactical concepts (making no reference to the meaning of the signs and the expressions), we can circumvent Wittgenstein's restriction (Carnap 1963a, 53–4).

A second philosophical breakthrough connected to the *Syntax* program came in October 1932, when Carnap, likely influenced by his insight that the protocol sentences debate turned not on questions of fact but on linguistic decision, realized (again, *pace* Wittgenstein) that *no* language is intrinsically correct – that there is no logical reality for a language to respond to. Sometime in 1933, most likely *after* his meetings with Quine, Carnap reformulated this insight as the *Principle of Tolerance* – the view that there are no morals in logic, and that "everyone is at liberty to build up his own logic, i.e. his own form of language, as he wishes" (Carnap 1934a, 52).

Carnap's innovations had important consequences for his views about philosophy. In the fall semester of 1932, Carnap finished what is now known as Part V of his *Logische Syntax* ("Philosophy and Syntax"), the chapter he was planning to discuss with the Vienna Circle before he fell ill.[25] In this chapter, Carnap develops a far-reaching new view about the nature of

---

[23] See, for example, Creath (1990c).
[24] See Uebel (2007b) for a more extensive discussion of this argument. Carus (2007b, 31) has a slightly different interpretation of Wittgenstein's argument, suggesting that we cannot meaningfully talk about the logical form of language in the *Tractarian* system because "statements about language" cannot "be construed as truth-functional concatenations of atomic sentences."
[25] See Carnap's letter to Schlick (November 28, 1932, RCP, 029-29-02).

philosophical questions, arguing that all problems of philosophy are logical questions and that all logical questions can be formulated as syntactical questions. Where Carnap, like many fellow members of the Vienna Circle, had previously denounced the "suppositious sentences of metaphysics, of the philosophy of values, of ethics" as devoid of cognitive content, he took a next step in the *Syntax* by arguing that the *remaining* questions of philosophy – for example, questions about "mankind, society, language, history, economics, nature, space and time, causality" – are only meaningful if they are reinterpreted as logical questions: "The supposed peculiarly philosophical point of view from which the objects of science are to be investigated proves to be illusory, just as, previously the supposed peculiarly philosophical realm of objects proper to metaphysics [e.g. the thing-in itself or the ultimate cause of the world] disappeared under analysis" (1934a, §72).

1.9 Carnap and Quine

Quine spent most of his days in Prague systematically studying Carnap's *Syntax* program. Not only was Carnap's logic course primarily concerned with his "new research on logical syntax,"[26] but Quine also extensively studied the manuscript in private and discussed his questions and comments with Carnap during their frequent and lengthy meetings. As a result of these discussions, Quine concluded that Carnap's *Syntax* offered novel solutions to a range of philosophical problems he had been struggling with himself.

Quine's correspondence and reports reveal that Carnap's *Syntax* program influenced his philosophical development in two ways. The first point Quine mentions in his report is that Carnap's book "answered" to his "satisfaction the question of the epistemological status of mathematics and logic," adding that this question was "formerly perplexing" to him (January 8, 1934, WVQP, Item 3254). Before his meetings with Carnap, we have seen, Quine accepted a holistic theory of knowledge that failed to account for logical and mathematical knowledge, except for the sketchy remark that the "eternal validity of logic" should be explained in (Lewisian) conventionalist terms (see Section 1.3). Carnap's book, Quine discovered, offered a conventionalist theory that did just this. In Carnap's system, one can simply *decide* to build logic and mathematics into the transformation rules of one's language; or, as Quine would put it two years later in his

---

[26] Carnap to Quine, February 6, 1933 (Quine and Carnap 1990, 110).

review of the *Logische Syntax*, in Carnap's system logic and mathematics acquire "apodictic validity through convention" (Quine 1935, 394).

The second (and most important) way in which Carnap influenced Quine's development is with respect to the status of philosophy. In his report, Quine explains that Carnap's "coming book ... has afforded the most satisfactory answer I have yet found to the still more perplexing question of the nature of non-meaningless philosophy" (January 8, 1934, WVQP, Item 3254). Quine, who, as we have seen, held conflicting views about the status of epistemology – vacillating between behaviorist and phenomenalist perspectives on knowledge – was clearly swayed by Carnap's theory that philosophy, too, is syntax:

> The way out of the jungle, Carnap ... claims, is through syntax ... all that is not meaningless in philosophy itself (this residue is, I should judge, mainly epistemology) speaks, when properly analyzed, not of things or 'reality' but rather of *syntax* ... Actually, when one reflects, this is the doctrine to which Lewis himself should logically have been driven. Lewis claims that all *a priori* truths are valid through definition ... Further, Lewis would certainly admit that epistemology or anything else in philosophy cannot be empirical, for then it would simply be natural science. Hence ... Lewis himself [would] be faced [with] the conclusion that philosophical truths are ... conventionally valid. (Quine to Cooley, April 4, 1933, WVQP, Item 260)

But that was not all. Lewis' theory, Quine maintained, was not only problematic because it did not offer a satisfying *view* about the nature of epistemology; it was also problematic because it was self-referentially inconsistent. Lewis had no satisfying answer to the question of how his philosophy could be justified considering his views about the nature of justification. Carnap's thesis that philosophy is syntax, on the other hand, had the benefit that it was self-referentially consistent:

> Every ... philosophy I know has the following difficulty. One reads the arguments of a given system of philosophy and perhaps agrees heartily throughout (this was my experience with Lewis' book), but at the end one remains with the problem of the status and the methods of the book which one has been reading, according to the philosophy set forth in that book itself ... How ... is the philosophy arrived at? Revelation, mysterious intuition, or arbitrary fiction? ... This whole bootstrap-tugging situation disappears in Carnap's view. He claims that philosophy is syntax; his claim is itself syntax and there is no circularity. (April 4, 1933, WVQP, Item 260)

Quine, in sum, was convinced that Carnap had solved one of the classical problems of especially empiricist philosophy, a problem that had prevented him from fully accepting the conclusions of Lewis' theory. Carnap,

Quine came to believe, had solved his questions about the nature of philosophy by showing (1) that all philosophy is syntax and (2) that the decision to view philosophy as syntax is itself a syntactical convention.

1.10 Carnap vs. Quine

Quine believed that Carnap had effectively solved some of the most "perplexing" questions of philosophy. In consequence, he began to see himself as Carnap's "disciple" for a number of years (Quine 1970a, 41), spreading the word about the latter's *Syntax* program through his teaching, via his writings (Quine 1935, 1936, 1937), and through his seminal "Lectures on Carnap" at Harvard in of 1934.

Still, it would be a mistake to suppose that Carnap's influence resolved the fundamental tension in Quine's philosophy. Quite the reverse. Although Quine believed that Carnap's *Syntax* program had solved his metaphilosophical qualms, his post-Prague papers show that he (unconsciously) imported *both* his behaviorist *and* his phenomenalist leanings into his interpretation of Carnap's framework. On the one hand, Quine seems to have already been giving a behaviorist spin to Carnap's program in the very weeks that he was studying the latter's manuscript in Prague. For one of the notes that *Carnap* wrote about his Prague discussions with Quine reveal that the latter was implicitly translating Carnap's distinction between the analytic and the synthetic into his own behavioristic epistemology:

> He said after reading my MS 'Syntax': 1. Is there a principled distinction between the logical laws and the empirical statements? He thinks not. Perhaps though it is only expedient, I seek a distinction, but it appears he is right: gradual difference: they are the sentences that we want to hold fast.[27]
> (March 31, 1933, RCP, 102-60-12)

Commentators have sometimes argued that Carnap's note shows that Quine was skeptical about the analytic–synthetic distinction from the very beginning. There may be some truth to this interpretation but the fact is that Quine kept searching for a valid way to draw the distinction until the late 1940s, when he realized that we do not need the distinction to account for our logical and mathematical knowledge. What is more important, I think, is that the note shows that Quine interpreted Carnap's theory through the lens of his own behavioristic epistemology, classifying the

---

[27] See also Tennant (1994).

truths of logic and mathematics as analytic because it is a *psychological* fact that we will not give them up in the light of adverse experience.[28] In "Truth by Convention," for instance, Quine argued that "the apparent contrast between logico-mathematical truths and others … [v]iewed behavioristically … retains reality as a contrast between more and less firmly accepted statements" and that this contrast "obtains *antecedently* to any *post facto* fashioning of conventions" (1936, 102, my emphasis).[29] Whereas Carnap was fundamentally committed to his principle of tolerance – accepting that we can decide which statements to build into the syntax of one's language – Quine's reinterpretation significantly diminished the relevance of Carnap's principle: we start with a system of accepted sentences and the only thing we get to decide is which of *these* statements we render analytic using Carnap's "technique of conventional truth assignment" (ibid.). The problem, however, is that Quine did not yet seem to realize that this was far from Carnap's way of characterizing conventionalism.[30]

This brings us to Quine's phenomenalism, the second of his conflicting philosophical commitments. In his "Lectures on Carnap," Quine makes it clear from the outset that he will discuss only "Carnap's very recent work" and exclude *Der Logische Aufbau der Welt* from his exposition. Still, we have seen that Quine was deeply impressed by the *Aufbau* and that he interpreted it in phenomenalist terms – that is, he viewed it as "carrying out in detail that to which Russell had merely pointed in his doctrine of logical constructions" (see Section 1.5). Quine's decision to omit the *Aufbau* from his lectures seems surprising because Quine's letter to Cooley reveals that Carnap and Quine also discussed the *Aufbau* during their meetings in Prague and that Carnap told him about some of the changes to his epistemology:

> Carnap has … departed in some fundamental respects from the point of view of the *Aufbau*; but in respects which, I believe, mark improvement. His departures turn in large measure upon a new opinion regarding

---

[28] See Verhaegh (2018, ch. 6). Perhaps Quine felt justified in his interpretation because he mistook Carnap's physicalism for behaviorism. See, e.g., Quine (1974b, 291): "Back in the 20s I had imbibed behaviorism at Oberlin from Raymond Stetson, who had wisely required us to study John B. Watson's *Psychology from the Standpoint of a Behaviorist*. In Czechoslovakia a few years later I had been confirmed in my behaviorism by Rudolf Carnap's physicalism." I thank Nathan Kirkwood for this suggestion.

[29] See also Hylton (2001), who rightly argues that passages like these "reveal fundamental assumptions" in Quine's philosophy "that are at odds with the views they espouse … about analyticity" (258). In particular, Hylton wonders how analyticity can play any explanatory role in Quine's philosophy of logic, considering the fact that we can only "post facto" impose a system of definitions which makes some of the sentences analytic.

[30] In Verhaegh (2018, §6.2.4), I argue that he only started to see this in 1943, when Quine realized that Carnap had a very different conception of language.

Carnap and Quine: First Encounters (1932–1936)

Protokollsätze. ... Of late ... Carnap (following O. Neurath) has come rather to the view that there is no ... stopping point but rather an indefinite or infinite regress, and that 'Protokollsätze' is merely a relative term. (Quine to Cooley, April 4, WVQP, Item 26)

It is clear why Quine, who had always rejected *naïve* sense-data theories himself (see Section 1.3), qualified Carnap's relativity thesis as an "improvement." What Quine failed to see, however, is that the *Syntax* program had *replaced* Carnap's rational reconstruction program. Again, the problem seems to be that Quine misinterpreted the radical nature of Carnap's principle of tolerance. Whereas Carnap believed that the question of what protocol language to adopt is a question of linguistic decision (see Section 1.5), Quine seems to have mistakenly presupposed that Carnap revised some of the details of his view about the nature of *protocol sentences* without abandoning the *Aufbau* program itself. Indeed, in the remainder of his letter to Cooley, Quine wrongly suggests that Carnap "would allow the Konstitution system to remain, but without claiming epistemological significance for the particular choice of primitive idea" and he apologizes that he "cannot explain exactly how the syntactic point of view" connects to the "epistemology and the relativity of Protokollsätze," as the connection is "not treated in his coming book" (ibid.). Quine, in sum, seems to have been unaware of the fact that Carnap had abandoned rational reconstruction for the logic of science;[31] it is for this reason, I think, that Quine's encounters with Carnap did not make him abandon his phenomenalism but kept the tension in his philosophy alive, though in a new Carnapian framework. It is also for this reason that Quine kept flirting with variants of phenomenalism, until he, in the early 1950s, finally came to see that one *can* develop a consistent behavioristic-naturalistic perspective if one replaces talk about sense data with talk about sensory stimulations and nerve endings.[32]

1.11 Epilogue

The Quines returned to the United States in June 1933, after they spent the remainder of the academic year in Poland, where Quine had the opportunity to present his work to the Lvov-Warsaw school and to discuss logic with Tarski, Lesniewski, and Lukasiewicz. Back in Cambridge, Quine would come to play a crucial role in promoting Carnap's *Syntax* program in the United States. For not only did he write a glowing review

---

[31] See Carnap (1936b).
[32] See Verhaegh (2018, ch. 5).

of Carnap's book in an American journal (Quine 1935), he also spread the word via his above-mentioned lectures about the *Logische Syntax*, which created a sustained interest in Carnap at Harvard. A few days after the third lecture, Quine wrote:

> I had a distinguished audience, comprising an assortment of professors and graduate students from many departments. ... The whole situation of the lectures was unique ... I stood under a bas-relief of the late metaphysician George Herbert Palmer, telling a gathering of professional philosophers that philosophy is nothing but syntax and that metaphysics is nonsense! ... [T]he attention was undivided. I have been meeting Professors Lewis and Sheffer weekly to discuss Carnap and be plied with questions; I am meeting them again this morning. So there is quite a stir about Carnap; a healthy phenomenon. (November 27, 1934, DBQ21)

Most importantly, Quine actively tried to arrange a position for Carnap in the United States. Carnap, who was becoming increasingly worried about the political developments in Central Europe, seems to have been considering emigrating to the United States for a number of years.[33] Quine's visit substantially sped up the process. Carnap's diary entries from March 1933 show that they spoke frequently about academic life in America and Carnap's prospects in the United States:

> March 4: "Quines with us ... Tell about ... America."
> March 22: "Afternoon 4-8 Quine here ... He says that in America most professionals ... are socialists."
> April 4: "Afternoon Quines here for the last time. They tell me, if it does not work out with Rockefeller, to write to American universities. They believe I certainly have prospects." (RCP, 025-75-11)

Between 1933 and 1934, Carnap tried to secure a one-year Rockefeller Fellowship and published a few papers in English (Carnap 1932a, 1934bc, 1935) in order to "naturally facilitate a professorship."[34] Quine, meanwhile, asked Whitehead, Sheffer, Lewis, and Huntington to write recommendation letters to the Rockefeller Foundation. In fact, Quine's lectures on Carnap were an important part of the campaign to get Carnap a professorship at Harvard. A few months before the lectures, Quine writes:

> Dr. Henderson, chairman of the Society of Fellows, and Professor Perry, chairman of the philosophy department of the university, seem to have got together on a plan to have me give a couple of lectures on Carnap's ideas.

---

[33] For a reconstruction, see Verhaegh (2020b).
[34] Carnap to Quine, June 4, 1933 (Quine and Carnap 1990, 120).

Carnap has for some time been anxious to teach in an American university, and during the past year I have taken all opportunities to push the matter with those in power here. ... Now I think there may be hidden motives behind their inviting me to speak on Carnap: ... more dope on Carnap as a possible Harvard professor. (September 29, 1934, DBQ21)

Unfortunately, it would take a few more years before Carnap could finally move to the United States. For it quickly became clear that there were "almost no available places in the whole country" due to the "economic situation."[35] In March 1935, however, Quine's propaganda started to pay off as the President of Harvard invited Carnap to Cambridge to receive an honorary degree – an invitation that would quickly turn into a lecture tour and that would eventually land him, with the help of Charles Morris, a position at the University of Chicago.[36]

Once in the United States, Carnap was reunited with his disciple. And although the philosophical tensions between Carnap's program and Quine's interpretation would start to surface a few years later (Verhaegh 2018, ch. 6), thereby triggering one of the most influential debates in the history of analytic philosophy, the philosophical friendship that had emerged in Prague in March 1933 would prove to be a stable one. For, as Quine would later write in his "Homage to Rudolf Carnap," even when they disagreed, Carnap was "setting the theme"; his philosophical development kept being determined by the problems he felt Carnap's position presented (Quine 1970a, 41).

Archival Sources

R. Carnap Papers. Archives of Scientific Philosophy. Hillman Library, University of Pittsburgh.
W. V. Quine Papers. MS Am 2587. Houghton Library, Harvard University.
W. V. Quine Unprocessed Papers. Private Collection, Douglas B. Quine.

---

[35] Quine to Carnap, March 12, 1934 (Quine and Carnap, 125–132).
[36] See Verhaegh (2020bc) for a reconstruction.

CHAPTER 2

On Quine's Guess about Neurath's Influence on Carnap's Aufbau

Thomas Uebel

2.1 Introduction

On two occasions of responding to papers about the relation of his philosophy to that of Rudolf Carnap, W. V. O. Quine hazarded a guess about the history of *Der logische Aufbau der Welt* (*The Logical Structure of the World*, hereafter *Aufbau*) that has rarely been discussed.[1] Referring, one gathers, to the passages of the *Aufbau* announcing that a system of concepts displaying their logical constitution could have been erected just as well on a physicalist basis instead of the phenomenalist one he did employ, Quine remarked:

> I picture Carnap as having been a single-minded phenomenalist when he devised the constructions that went into the *Aufbau*. When the book was ready for printing, I picture Neurath pressing the claims of physicalism. I then picture Carnap writing and inserting those paragraphs of disavowal by way of reconciling the book with his changing views. Significantly, he took the physicalist line in his subsequent writings, and refused permission to translate the *Aufbau* for more than thirty years. (1994a, 345; cf. 1990a, 67)

This remark is typically read, I suspect, in light of Quine's well-known and of late much-criticized reading of the *Aufbau* as a foundationalist tract in the tradition of Bertrand Russell's external world program – and possibly dismissed for this reason.[2] What is not considered is whether this remark, once framed differently, does not after all capture a true dynamic and so puts the finger on the beginnings of the *Aufbau*'s undoing. Such a suggestion may well be met with incredulity. "Why be a 'single-minded phenomenalist' if not to pursue a foundationalist agenda?" it will be

---

[1] All references in the text and footnotes solely by "*A*" are to this book: Carnap (1928a/1967). Quine's responses were to Creath (1990a) and Tennant (1994).

[2] See, e.g., Quine (1951f) and (1969a). For criticisms see, e.g., Friedman (1987) and (1992) and Richardson (1998). But note Quine's disavowal of Carnap's concern with certainty at (1995a, 13).

asked, supposing that, surely, Carnap will not have been moved by idealist yearnings.

The question is a good one. To answer it we must, first, specify what shall be meant here by "the *Aufbau*." Meant by the definite article (I follow Quine's usage here) is the particular system of logical construction or "constitution" of the objects of cognition that Carnap did develop in the book, not the other systems noted as possible in the "paragraphs of disavowal." A central assumption in the system that Carnap did develop (but not in the others) concerns the order of epistemic priority: "The autopsychological objects are epistemically primary relative to the physical objects, while the heteropsychological objects are secondary. Hence we shall constitute the physical objects from the autopsychological ones and the heteropsychological from the physical objects" (§58, 94). This is Carnap's doctrine of "methodological solipsism."

Then we must clarify what Quine surely understood as well, namely that ontology was not at issue, but only "methodological" phenomenalism. So the supposed "single-mindedness" also needs explanation. I take it to refer to the fact that none of the logical constructions of objects attempted in the *Aufbau* transcend the phenomenal consciousness of an individual subject.[3] The physical objects, other minds and cultural objects which Carnap reconstructed are but simulacra of their real counterparts which fall away as dispensable for epistemological purposes once his logical constructions are available. (No foundationalism needed here.) Quine's guess has it then that, but for Neurath's insistence, the paragraphs asserting that other types of constitutions of conceptual systems are possible would have been missing.

What may prompt skepticism about Quine's claim is that current understandings of the *Aufbau*'s aim contradict the foundationalist interpretation Quine associated with its "single-minded" phenomenalism. Rather, apart from seeking to demonstrate the unity of science by exhibiting in outline how all concepts can be generated from the same basis, Carnap's aim was demonstrating that "even though the subjective origin of all knowledge lies in the contents of experiences and their connections, it is still possible ... to advance to an intersubjective, objective world, which can be conceptually comprehended and which is identical for all observers" (*A*, §2, 7). A successful simulation of human knowledge under carefully controlled

---

[3] Thus Carnap stated that "*on no level* of the constitutional system, hence not even through the utilization of the reports of other persons, *is something fundamentally new introduced into the system*, but that what we have here is *only a reorganization* (albeit a very complicated one) of the given elements" (*A*, §144, 222, orig. emphasis) and that even concerning the "world of the other," it remains the case that "*we do not desert the autopsychological basis*" (*A* §145, 223, orig. emphasis).

thought-experimental conditions was to show that what matters for objectivity is not the relation knowledge claims bear to what they are about – though that matters for their truth – but that their content is expressible in purely structural terms, without reference to intersubjectively inaccessible manifestations of subjectivity such as intuition or to meanings seemingly reaching out beyond experience itself.[4] However, far from rendering Quine's guess implausible, this structuralist reading affords a perspective able to sustain it. For it turns out that attributing the structuralist program is no hindrance to attributing central importance to methodological solipsism in the *Aufbau* – and objecting to its deployment if so inclined. After all, the structuralist program too requires the radical reduction of all concepts to logical permutations and iterations of just one basic relation (remembered similarity) that Carnap hoped to complete by excising the need to appeal to the meaning even of the basic relation itself.[5] To render content specifiable in purely structural terms demands that there is no need for reference or any other relation to objects apart from or outside of the structure at issue.

Yet there are other reasons to question the accuracy of Quine's guess if it were read to indicate much more than the general direction of Carnap's development or even Neurath's prescient opposition. To be sure, both ended up as physicalists of one variety or other, but that convergence was hard won and not achieved by argument along identical lines, and, moreover, was not achieved at the early stage at issue. (Their part in the Circle's notorious debate over the content, linguistic form and epistemological status of so-called protocol sentences still lay years ahead.) Of course, that "those paragraphs of disavowal" would have made for substantial additions of text (four paragraphs of six in §57, all six paragraphs of §59, all five of §62) and integrative reformulations (the first two paragraphs of §64) does not speak against his guess, for these changes would have remained clearly delimited. That Carnap's "profession of neutrality between a phenomenalistic basis and a physicalistic one" (Quine 1994a, 345) should have been a relatively recent addition has a distinct plausibility. Yet is it the case that "Neurath talked him around to physicalism before the book was finished" (1990a, 67)? This can only be decided once we can see what precisely Carnap committed himself to in these paragraphs. As we will see, the matter of physicalism is far from straightforward.

---

[4] See fn. 2 and add Ricketts (2010) to the critics; the Friedman reading is also partially endorsed in Pincock (2005) and Uebel (2007a, ch. 2).

[5] The latter move already worried Carnap (*A* §§153–155) and remembered as the "foundedness problem" is now widely held to have failed: see, e.g., Friedman (1987) but note MacBride (2021, §§5–6).

On Quine's Guess

But why, it may be asked, focus on Quine's remark? The reason is that it directs attention to the convergence on physicalism by Carnap, Neurath and Quine. This convergence is highly significant for it highlights the rejection of the epistemological "given" by the former two standard-bearers of logical positivism, a rejection achieved long before it became the fashionable battle cry of post-positivism.[6] Different stages of this development deserve closer investigation; here I investigate its perhaps earliest stage from the vantage point of Quine's guess.[7] After a brisk survey of the *Aufbau*'s pre-publication history and of Neurath's prompt review of it, I consider the significance of a small note of Carnap's concerning the *Aufbau* in the light of his diary entry about pre-publication discussions of the book manuscript with Neurath.

2.2 The *Aufbau* in the Circle and Neurath's Review

In January 1925 Carnap went to Vienna at the invitation of Schlick, who previously had agreed to support his *Habilitation*, and gave two talks to his discussion group. The first was prearranged and concerned the topology of space-time, the second was added to present his research project. In his autobiography Carnap reported: "From the beginning, when in 1925 I explained in the Circle the general plan and method of *Der logische Aufbau*, I found a lively interest. When I returned to Vienna in 1926, the typescript of the first version of the book was read by members of the Circle, and many of its problems were thoroughly discussed" (1963a, 20). The further development of the *Aufbau* project took place in two stages: the first one up to the submission of his manuscript for the *Habilitation* in December 1925, and the second one, of revising and radically shortening that text for submission to the printers, in January 1928.[8] Schlick had studied Carnap's typescript by early March 1926, after which it began making the rounds of members of his discussion group, beginning with Waismann.[9] Unfortunately, no copy of it, generally

---

[6] Not untypically, followers of Wilfrid Sellars' deconstruction of "the myth of the given" (1956) pin such givenist mythmaking on the logical positivists, as Richard Rorty did in his preface to Robert Brandom's 1997 edition of Sellars' seminal paper. That Quine remained not wholly innocent in this regard must also be noted.

[7] For discussion of Carnap's and Neurath's path to the rejection of methodological solipsism, see Uebel (2021). The present paper builds on parts of Uebel (2016) with a different focus and improved argument.

[8] See Carnap to Schlick, 2 and 11 December 1925 and Carnap's diary for December 1925 and 27 January 1928 (RC 029-32-33, 029-32-32, 25-72-04 and 025-72-02 ASP).

[9] See Schlick to Carnap, 7 and 14 March 1926 (RC 029-32-27 and 029-32-17 ASP).

referred to as "Konstitutionstheorie" (Constitution Theory), appears to have remained in existence and it requires circumstantial reasoning to establish the changes it underwent to become the *Aufbau* we know.[10] It is into the second period that Neurath's pre-publication criticism falls. But first let us consider Neurath's review of the *Aufbau* and the simultaneously published *Scheinprobleme der Philosophie* in the Austro-Marxist monthly *Der Kampf* in the fall of 1928 (apparently the first review the book received).[11]

Exhorting its readers to study both books reviewed and, more generally, the "empirical rationalism" of the "'Vienna school' around Moritz Schlick" so as to render it "useful for Marxism" (1928, 296–297), Neurath introduced Carnap's *Aufbau* as an attempt "to characterize completely systematically and comprehensively the foundations of exact empirical knowledge."

> Carnap seeks to show how to arrive at a consistent view of the world once we discount all accidental and variable sense impressions. He undertakes to characterize sense impressions on the basis of certain order-structures, order-structures in which "red", "hard", "loud", "cis" etc. do *not* appear, but only facts which can be captured by mathematical-logical means – *and that suffices*! Carnap consciously turns away from taking empathy in any form, or personal attitudes, as his starting point. He only knows that kind of insight which can be grasped by every human being! Structural order is what is most common, what is most universal in our experience of things! (Ibid., 296, orig. emphasis)

What struck Neurath was Carnap's abstraction from all subjective elements of experience, his thoroughgoing objectification-by-structuralization of knowledge by the logico-mathematical means deployed. Given his own long-standing opposition to the categorical separation of the *Geistes-* from the *Naturwissenschaften* (e.g., 1910, 267), Neurath found Carnap's project in the *Aufbau* very congenial. One of Carnap's purposes was, after all, to establish that all the sciences shared the same constructional system of concepts; this provided valuable support for Neurath's cherished idea of a unified science. But it is also of great significance that Neurath appreciated the point of Carnap's structuralist methodology: the centrality of

---

[10] Besides the original sketch "Von Chaos zur Wirklichkeit" of July 1922, there exists a three-page plan "Entwurf einer Konstitutionstheorie der Erkenntnisgegenstände" dating from January 1925 and the notes for two of three lectures given in June and early July 1926 in Vienna, titled "Thesen zur Konstitutionstheorie" (RC 081-05-01, 081-05-02 and 081-05-07 ASP, respectively). On these early stages of Carnap's project, see Carus (2016).
[11] Compare the list in Benson (1963, 1059).

intersubjectivity to all scientific knowledge. Thus he stressed that Carnap "only knows that kind of insight which can be grasped by every human being," evidently attempting to pre-empt the potential dismissal of Carnap's methodological solipsism by the "comrades" as a bourgeois Robinson Crusoe fantasy.

Methodological solipsism is not mentioned as such in Neurath's review, but subtle criticism of it can be detected. It lies behind a dense passage which, on the face of it, only objects to apparent anticipation of an "ideal language" and "complete insight" (1928, 296). There Neurath stressed what Carnap would have been the first to concede, that his *Aufbau* represented a highly idealized picture of human knowledge – and here Neurath projected Carnap's project well beyond what it actually showed – that was fictitiously complete and definite. More specifically, Neurath insisted that theory choice does not "flow from the subject matter itself" (ibid.). What, for Neurath, were at least partially conventionally and so socially determined choices in theory construction were rendered invisible, in the *Aufbau*, by the idealizing assumption that the rational reconstructor knew what must be the end result of constitution theory and determined the reductive definitions accordingly. In the *Aufbau*, knowledge "flowed from" its subject matter only because the results of historically prior theory choice were appropriated in this way. What lies at the bottom of Neurath's criticism of the *Aufbau* then were suspicions about the philosophical assumptions packed into the position of methodological solipsism: the idea of the epistemic self-sufficiency of a solitary individual. To say so explicitly, of course, would have been to accuse Carnap precisely of what Neurath sought to protect him from, so he recast Carnap's rational reconstruction as a philosopher's anticipation of future science.[12]

2.3 Neurath's Pre-Publication Criticism

In the light of this discrete but fundamental criticism, it is of interest to consider a little document found in Carnap's *Nachlass* called "Neurath über Konstit[tuitions]theorie." Written in Carnap's shorthand and dated "21. 11. 26," it records what appears to have been Neurath's comments about the circulating typescript of what Carnap later called "the first version" of the *Aufbau*.

---

[12] Neurath's review is discussed in greater detail and context in Uebel (2007a, 105–112).

The most relevant parts read as follows:

> "My exposition is turned unfortunately more against realism than idealism. Too much emphasis on methodological solipsism. That sounds too individualistic. Emphasize more the '*objectivism*'. *Say right at the start that the goal is an objective world, the same for all individuals.* El[ucidate].
> §224. *The realism of the physicist remains intact*, but only will be corrected in the direction of objectivism. Perhaps as follows: the lawful connections are objective, i.e., do not depend on the will of the individual; but there is no matter to which 'reality' could be ascribed; that is a metaphysical concept."[13]

Neurath also made further presentational suggestions and commented favorably on the holistic nature of the constitution of physical space. Of interest here is his concern that it should be plain and evident to the reader that the construction system of the *Aufbau* terminated in the concepts of an objective physical world.

Now, did Neurath's criticism of the overemphasis on methodological solipsism mean that Carnap's work merely *sounded* too idealistic or that it still *was* too idealistic? It does not seem to be the case that already at this stage Neurath had to hand the convincing argument to the effect that any suitable comprehension of our own experience requires that we think in terms of an intersubjective world – the fulcrum of Wittgenstein's later *Philosophical Investigations* and Quine's *Word and Object* that Neurath anticipated in 1931.[14] However, already his review of 1928 criticized the epistemological conception that informed the *Aufbau* as too "individualistic" – a term also used in Carnap's note about Neurath's 1926 in-person criticism. This suggests that the latter also charged that the social element in theory choice and construction was unduly neglected.

Notably, while the *Aufbau* was in print in late February/early March 1928, Carnap sent Neurath a copy of an enclosure in a letter to Schlick from the previous December in which he had raised the question of its title and indicated that he had planned a later study of a constitution system with a physical basis.[15] Carnap remarked to Neurath that he sent it to him "since you will have particular interest in the conceptual system with a physical

---

[13] So recorded in Carnap's shorthand (with his underlining here given as italics): "Neurath über Konsti[tutions]theorie," 21 November 1926 (RC 029-19-04 ASP, transcription by Brigitte Parakenings). The *Aufbau* does not contain a §224, of course, as it was much shortened for publication.

[14] For discussion of repeated employments of Neurath's private language argument, see Uebel (2007a, 226–252) and (2021).

[15] Manninen (2002, fn. 14) argues persuasively that it was a copy of the enclosure "Frage über die Wahl des Buchtitels" of the letter Carnap sent to Schlick, December 23, 1927. The enclosure of the letter to Neurath has not been preserved.

basis. I take it that we will be in agreement concerning the advantage, but also the disadvantage of this system in comparison with that on an autopsychological basis."[16] This suggests that Carnap and Neurath had previously discussed the relative advantages of these systems. Carnap implied ("I take it") that Neurath did not as yet dispute his claim that the physical constitution system had the disadvantage that its "ordering of objects" was not "a correct reflection of the epistemic relation" (*A* §59, 96). Whether Carnap was correct on this point may be doubted: Neurath did not yet have a knock-down argument against methodological solipsism, but he already was deeply suspicious of it, as we saw.

Now the advantage of the physical constitution system was described in the *Aufbau* thus: "[I]t uses as its basic domain the only domain (namely, the physical) which is characterized by a clear regularity of its process. In this system form, psychological and cultural events become dependent upon the physical objects because of the way they are constructed. Thus they are placed within the one law-governed total process" (*A* §59, 95).

Carnap concluded that, "from the standpoint of empirical science," the system with a physical basis provides "a more appropriate arrangement of concepts than any other" because "the task of empirical science (natural science, psychology, cultural science) consists, on the one hand, in the discovery of general laws, and, on the other hand, in the explanation of individual events through their subsumption under general laws" (ibid.) This advantage of the system with a physical basis is clearly related to the "objectivism" that Neurath had been urging in 1926 – the subject-independence of the physical realm. Yet Carnap did not elaborate and in the *Aufbau* only touched on this point once more later on: the autopsychological lacks a thoroughgoing regularity because it is not closed causally (*A* §132, 204).

As we shall see, this was by no means a new idea of Carnap's, but it is also clear that in their discussions Carnap and Neurath raised the idea of an alternative constitution system to the one actually chosen. Whether these discussions were the only impetus for Carnap's "paragraphs of disavowal" may be doubted therefore, but to assess the extent of Neurath's influence we must also investigate whether the "neutrality" Carnap avowed was one between (methodological) phenomenalism and physicalism, in a sense of "physicalism" that is commensurate with that of his well-known papers of 1932, as Quine's remark suggests.[17]

---

[16] Carnap to Neurath, 25 February–3 March 1928 (RC 029-16-05 ASP).
[17] See Carnap (1932a) and (1932b). For criticism of even these as not yet radical enough, see Neurath (1932) and Carnap's response (1932c).

2.4 The *Aufbau* and Physicalism

It may be wondered whether it makes sense to associate physicalism with the *Aufbau* at all – was it not a phenomenalist undertaking? The answer directs us to the section of the *Aufbau* already quoted from: §59, entitled "A system form with a physical basis." It concludes by stating that "science ... needs both an experiential and a materialistic derivation of all concepts" (96). So the *Aufbau* did recognize both the possibility and a certain necessity of a materialistic constitution system.

We must ask: what is the nature of this "need"? And why can "a more detailed description of [the materialistic] system and its importance for science" not be given "at this time," as Carnap claimed (ibid., 95)? Granting correctly that what makes a materialistic constructional system more "appropriate" for empirical science is that its "order of construction" reflects the causal order of the "law-governed total process" (just as the system with an autopsychological basis reflects the presumed epistemic order) leaves the second question unanswered. In §62 Carnap presented clear proposals for three kinds of possible physical bases for a constitution system: was it simply lack of space and time that prevented Carnap from providing a more detailed characterization of these systems?

More radically, we must ask: Would even the mere availability of a constitution system with a physical basis be sufficient grounds to speak of physicalism in the *Aufbau*? Consider what Carnap called "physicalism" in 1932: the thesis that the statements of the languages of all the sciences are translatable into the language of physics.[18] So physicalism, for Carnap, was, first of all, a *metalinguistic* thesis that, second, declared the *primacy of the physical language*. Neither of these characteristics are satisfied by the possibility or actuality of a materialistic constitution system that Carnap described on the *Aufbau*.

That noted, let us consider whether what might be regarded as a precursor of physicalism can be found in the *Aufbau*. Carnap confidently wrote that "physical objects are reducible to psychological objects and vice versa" (title of §57) and spoke of "their mutual reducibility" (§58, 93). On pain of being "suspended in the void" epistemologically, "statements about physical objects can be transformed into statements about perceptions (i.e., psychological objects)" and "every statement about a psychological object can be transformed into a statement about those indicators" from which it is "inferred" (§57, 92–93). So while in the *Aufbau* Carnap did not as yet

---

[18] See Carnap (1932a, 67).

defend an explicitly metalinguistic thesis or consider one object domain to be primary, he did announce the intertranslatability of talk of psychological and physical objects. Not yet drawing the distinction between object- and metalanguage, Carnap intended his reduction of object types to be equivalent to the reduction of the concepts corresponding to them: object reduction amounted to linguistic translatability.[19] The question is how far this "mutual reducibility" can take us. Does it amount to a premetalinguistic version of physicalism that also abjures primacy claims?

There are two readings of what we can call the "intertranslatability thesis," a weak and a strong one. The *weak version* simply says that (i) there exist two constitutional systems, one taking elements and relations of the physical domain as basic and one taking elements and relations of the psychological domain as basic, such that (ii) in the system with a physical basis the reduction of constructed psychological objects to physical ones and in the system with a psychological basis the reduction of constructed physical objects to psychological ones is effected. Moreover, (iii) there exists a recursive procedure for each constitution system to furnish statements that are extensionally equivalent to statements of the other (they have the same truth value). In consequence of (i)–(iii), neither of the two systems can be held to possess overall primacy.[20]

The *strong version* of the intertranslatability thesis derives from Carnap's assertion of the mutual reducibility of physical and psychological objects as in (i)–(iii), and also the further claim that (iv) the basic language of the system with a physical basis and the basic language of the system with a psychological basis are mutually translatable. It then follows that (v) for all statements formulatable in one system an extensionally equivalent one can be formulated in the other. The strong and the weak versions are distinguished therefore by whether the extensional equivalences of *all* statements are held to be formulatable or not. The first thing to note then is that physicalism in the *Aufbau* even without primacy would require strong intertranslatability.

Yet which of the two readings of the intertranslatability thesis is appropriate for interpreting Carnap's *Aufbau* as it was written? Two further facts are relevant here. The first is that Carnap distinguished between two types of constructional systems with a psychological basis: one with "autopsychological objects" as its fundamental domain (first-person experiences)

---

[19] Carnap stated that "the word 'object' is here always used in the widest possible sense, namely, for anything about which a statement can be made" and that "it makes no logical difference whether a given sign denotes the concept or the object, or whether a sentence holds for objects or concepts" (*A* §5, 10).
[20] In this sense the intertranslatability thesis was invoked for expository purposes in Uebel (2007a, 38). The distinction between the weak and the strong versions was not drawn there, however.

and one with "heteropsychological objects" (other minds) as its fundamental domain (*A* §58). Physicalism in the *Aufbau* without primacy, strong intertranslatability, would require the mutual translatability of the language speaking of physical objects with the languages speaking of either type of psychological object. The second relevant fact is that in the *Aufbau* the autopsychological and the heteropsychological objects play different roles in relation to the physical objects. The heteropsychological objects are reducible to physical objects while the latter in turn are reducible to autopsychological objects. So it is not the same type of psychological object that is both reducible to and constitutive of physical objects. This asymmetry in the reduction relations between the physical objects and the two types of psychological objects is significant.

Strong intertranslatability takes Carnap's remark that physical and psychological objects are "mutually reducible" to suggest that the languages basic to the two constitution systems are fully intertranslatable. This suggestion trades, however, on treating "the psychological" as interchangeable under the different guises of first-person and third-person mental attributions, as indeed we do in ordinary parlance. Yet Carnap did not offer even an outline of the reduction of autopsychological objects to physical objects anywhere in the *Aufbau*. So the strong intertranslatability thesis suggested by Carnap's bold statement ("all psychological objects ...") disappears under analysis as a misleading and ultimately false conceptualization of what the *Aufbau* provided, for the psychological pole of the strong intertranslatability thesis is not univocal. After all, it is not the language which speaks of autopsychological objects that is reducible to the language speaking of physical objects and it is not the language that speaks of heteropsychological objects that the latter reduces to.

To see the failure of strong intertranslatability in greater detail, consider that Carnap outlined two routes of reduction of psychological objects to physical ones, one readily traveled every day and one open to us only in principle. The former turns on our reliance, for interpersonal understanding, on the expression and reporting relations holding between mental states and behavior (*A* §57, 93). This type of reduction was sharply contrasted with the reduction of psychological objects to physical ones on account of the presumed fact of psycho-physical mind–brain parallelism: due to "the present state of science" this was recognized to be possible only "in principle" (ibid., 92). Intertranslatability was demonstrable therefore only in the case of the former type of reduction, not in that of the latter. Yet even the former route was barred in the *Aufbau* if it was autopsychological statements one started with.

To be sure, Carnap wrote that "every heteropsychological process is in principle recognizable, that is, it can either be inferred from expressive motions or else questions can be asked about it." And, indeed, he continued: "Thus every statement about a psychological object can be transformed into a statement about those indicators. Thus it follows that all psychological objects can be reduced to expressive motions (in the wider sense), i.e., to physical objects" (93). Note, however, the easily overlooked *non sequitur* (the first "thus"): that *every* statement about a psychological object can be transformed into a statement about the object's indicators does *not* follow from the fact, granted here, that every statement about a heteropsychological object can be transformed into a statement about the behavioral indicators. For that it is also required that autopsychological statements can be so transformed – but precisely this we have not been shown. This was not an oversight, for it could not have been shown.

The *Aufbau* system reflected the (presumed) order of epistemic priority. It offered a reduction of understanding, not of things, and in this order autopsychological statements were primitive. Accordingly, the very idea of translating an autopsychological into a physical statement was nonsensical: it literally defied understanding. That the *Aufbau* did recognize a physical language, namely the one that translated heteropsychological statements, is beside the point. A translation of the autopsychological language could only be effected in a constitution system with a different order of priority. We must conclude that talk of "mutual reducibility" promises more than the *Aufbau* can deliver.

Strong intertranslatability cannot be sustained because (iv) is not supported: the basic language of the constitution system with an autopsychological base was not shown to be translatable into the basic language of the constitution system with a physical base. To be sure, weak intertranslatability, according to which two radically different constitution systems may nevertheless formulate some pairs of extensionally equivalent statements, still holds. But the important point is that in the *Aufbau* Carnap went no further for (v) is also false: it is not the case that for *every* statement formulatable in the constitution system with an autopsychological base, an extensionally equivalent one can be formulated in the constitution system with a physical base. Without support for strong intertranslatability the suggestion that the *Aufbau* affords a pre-metalinguistic version of Carnap's physicalism without claims to primacy stands refuted.

It is not the case then that Carnap simply made a promissory remark that he did not redeem but could have. First, note that his argumentation did show that the "mutual reducibility" of all psychological and physical objects, strong intertranslatability, obtains in principle – albeit on a condition that his *Aufbau* did not satisfy, namely, the provision of two distinct constitution systems, a physical one in addition to the methodologically solipsist one. Second, note how he conceived of the reach of the physical constitution systems that he outlined but did not elaborate further: "After we have constructed the physical objects by proceeding from such a physical basis, we can construct the other object types according to our earlier considerations concerning the reducibility of psychological objects to physical ones and of cultural objects to psychological ones" (§62, 100). That autopsychological ones are meant to be included here is unlikely for Carnap is best read here ("our earlier considerations") as speaking only of the reductions in the higher reaches of the *Aufbau*'s constitution system and leaving aside the contentious reduction of the autopsychological. Of course, this type of physicalist reduction – one which the *Aufbau* can sustain – amounts to weak intertranslatability and does not allow for unrestricted "mutual reducibility."

The resultant reading of the *Aufbau*, however unorthodox, is not arbitrary at all. It connects well with what Carnap's documented views in the years leading up to his physicalism papers of 1932. In these years too Carnap was not prepared to consider autopsychological statements translatable into physical statements, even though he was happy so to regard heteropsychological statements, because he held the former type of physicalist translation to be highly problematic. In unpublished drafts from summer 1930 for his "Die physikalische Sprache als Universalsprache Wissenschaft" (*Unity of Science*, 1932a) and "Psychology in physikalischer Spache" ("Psychology in Physicalist Language," 1932b), as well as in discussions of the Circle after he had already adopted a metalinguistic standpoint in 1930 and 1931, Carnap expressed views still strongly at variance with those propagated in the published versions of these papers.[21] There he held precisely that the autopsychological statements resisted translation into physicalist statements because some of their non-negligible content was lost thereby: "[A] sentence with which a subject speaks about an autopsychological process has a different meaning for a hearer than for a speaker."[22] Characteristically for this period, Carnap entertained a *dualism* of universal languages (languages able to translate all other languages) where, however, the physical language had its universality limited to languages expressing states of affairs that are "intersubjectively recognizable."[23]

---

[21] For documentation and discussion of these worries, see Uebel (2007a, 191–200 and 217–222). The quotations following in the text below are there backed up by further ones stemming from a lecture given in January 1931.

[22] Carnap, RC 110-03-36 ASP, 55.

[23] Carnap, RC 110-03-22 ASP, 20; cf. ibid., 24.

The original domain of the autopsychological language was excluded from its reach (as it was in the *Aufbau*).

With no reasons to think that Carnap's worries about physicalist translations of autopsychological statements in 1930 and 1931 were new ones, we can see reflected in them the lacuna of the *Aufbau* just documented remaining in place. Carnap's own objections to the full intertranslatability of the physical and the autopsychological languages were overcome only in the winter of 1931/32 by his introduction of the distinction between the formal and the material mode of speech. As he once put it to Neurath: "Only due to the sharp distinction and the rejection of the material mode has the elimination of the dualism of the two languages become possible."[24] The position reached thereby, of course, is Carnap's physicalism of *Unity of Science* and "Psychology in Physicalist Language." Because of what distinguishes this position, his old position in the *Aufbau*, lacking strong intertranslatability, cannot be considered an anticipation of fully fledged physicalism, albeit non-metalinguistic and without primacy claims. That Carnap stated "Since every scientific cognition can be formulated in physical language, we also call this 'physicalism'" in a lecture of January 1931 does not alter this fact, for it retained the restriction of the universality of the physical language to intersubjectively available states of affairs.[25]

Importantly, this is not to say that there is no continuity at all – and Carnap's from our perspective somewhat premature use of the term "physicalism" in 1931 points in this direction. Carnap's paragraphs of disavowal in the *Aufbau* amount to a pre-metalinguistic anticipation of Carnap's "physicalist" position in the 1930 drafts which, however, still abjured not only primacy claims over non-physicalist languages but also restricted the claim to universality. It is the latter type of restriction that carried over from the *Aufbau*, but now was paired with a de facto recognition of the physical language as equally basic to the phenomenal one. What is clear is that this restriction also decisively invalidates Quine's speculative guess about physicalism in Carnap's *Aufbau*.[26]

---

[24] Carnap to Neurath, 2 March 1932, RC 029-12-60/61, 2. The preceding sentence makes clear that what the dualism is of is one of the physical language and the autopsychological language. What makes their co-existence a dualism is the failure of strong intertranslatability.

[25] See Carnap, RC-081-03-85 ASP, 5. Notably, the term "physicalism" was not yet used in Carnap's 1930 drafts; its first documented use appears to be Neurath's, in correspondence in December 1930 (see Uebel 2007a, 182–183).

[26] Although Carnap had not advertised it in the *Aufbau*, already in the Circle Edgar Zilsel had remarked on the asymmetry of translatability into the physicalist language of the two psychological languages of the *Aufbau* (1932, 145–146). Related diagnoses are suggested but not further elaborated in Feigl (1950/1981, 289), Kim (2003, 269), and Ryckman (2007, 95).

2.5 What Neurath's Criticism Accomplished

So Carnap's *Aufbau* does not contain an early version of what we would recognize as physicalism at all but only an idiosyncratic version of empirical psycho-physical parallelism.[27] To be sure, Carnap's was pretty sophisticated in resolutely resisting any philosophical "interpretation" of it.[28] But does this failure to count as properly physicalistic mean there was no influence of Neurath's criticism on the *Aufbau* at all? To conclude this would be too quick. Did Neurath perhaps prompt what we could call Carnap's "proto-physicalism" (non-meta-linguistic and abjuring primacy and universality)?

Recall Neurath's comment on "§224": "*The realism of the physicist remains intact*, but only will be corrected in the direction of objectivism. Perhaps as follows: the lawful connections are objective, i.e., do not depend on the will of the individual; but there is no matter to which 'reality' could be ascribed; that is a metaphysical concept" (RC 029-19-04). Now interestingly, on the back of the very page on which Carnap noted Neurath's comment, Carnap wrote (again in shorthand):

> Δ to p. 566. It is occasionally said that there is a tacit realism at the bottom of the practical procedures of the empirical sciences, especially of physics. However, we must here clearly distinguish between a certain kind of language use and the assertion of a thesis. The realistic orientation of the physicist shows itself primarily in the use of realistic language; this is practical and justifiable (cf. §52). On the other hand, realism, as an explicit thesis, goes beyond this and is not permissible; it must be corrected so as to become 'objectivism': the regular connections (which in laws are formulated as implication statements) are objective and are independent of the will of the individual; on the other hand, the ascription of the property 'real' to any substance (be it matter, energy, electromagnetic field, or whatever) cannot be derived from any experience and hence would be metaphysical. (RC 029-19-04 verso)

This jotting is virtually identical to the third paragraph of §178 of the published *Aufbau*.[29] In fact, that here on the verso side Carnap even used

---

[27] For a discussion of the wide acceptance of psycho-physical parallelism as an empirical hypothesis in nineteenth- and early twentieth-century Germany and Austria – and of the different philosophical interpretations it was often given (all of which Carnap rejected) – see Heidelberger (2003).

[28] When Carnap held that "all types of psychological processes have physical parallels (in the central nervous system)" (*A* §57, 93), he endorsed psycho-physical parallelism purely as an empirical hypothesis and rejected the demand to "interpret" (not "explain" as George's translation erroneously has it) the correlation that makes for the empirical thesis of psycho-physical parallelism: "[T]he quest for an interpretation of that parallelism belongs within metaphysics" (*A* §169, 271, trans. amended).

[29] The two minor differences are that the published version has "(usually tacit)" instead of "tacit" and emphasis also on "so as to become 'objectivism'." (Needless to say, "Δ to p. 566" is missing there too.)

a paragraph numbering for a cross-reference (§52) that is correct for the published version – unlike his report of Neurath's comments from 21 November 1926 on the recto side (which referred to a §224 which does not exist in the published version) – strongly suggests that Carnap had gone back to his old notes at a later date to formulate more precisely the point they had originally discussed for inclusion in the final manuscript. That it does so, moreover, suggests to me also that the original point was Neurath's.[30] (The "objectivism" that Neurath defended represents a deflationist criticism of realism to be set alongside his more strident criticisms of idealist and phenomenalist metaphysics.)

Note also that Neurath's suggestion "say right at the start that the goal is an objective world, the same for all individuals" found a reflection at the end of §2 in the published *Aufbau*: "Even though the subjective origin of all knowledge lies in the contents of experiences and their connections, it is still possible, as the constructional system will show, to advance to an intersubjective, objective world, which can be conceptually comprehended and which is identical for all observers" (7). There would have been no reason for Neurath to have made his remark pre-publication (as he did) if the circulating draft had contained this sentence already.

So Neurath's influence on Carnap's *Aufbau* can be traced. His concern about how the *Aufbau*-to-be might be read found its author receptive and his response congenial. But again we must also ask whether Neurath's pre-publication criticism pertained to the substance or just the presentation of Carnap's views. For instance, already in Carnap's 1922 manuscript "Vom Chaos zur Wirklichkeit" – the very beginning of the *Aufbau* project – and in "Three-Dimensionality of Space and Causality" we find formulations of the *Aufbau*'s conception of the lawlessness of the psychological domains.[31] So the need for a physicalist constitution system to represent law-governed reality was not a new insight of the *Aufbau*.

Yet there was one aspect of Neurath's criticism in his review which went beyond anything anticipated by Carnap: that the *Aufbau*'s reconstruction of human knowledge overlooked the socially conditioned nature of conventionalist theory choice and development – a point which Neurath had pressed already in 1926. Carnap's diary entry for 21 November 1926 specifies his criticism more sharply than the note to himself so far discussed.

---

[30] Juha Manninen also wondered whether the passage in §178 was "new with respect to the original manuscript" and "how much of the passage … belongs to Neurath and how much to Carnap" (2003, 136–137).

[31] See RC 081-05-01 ASP, 12–13, and Carnap (1924, 123). According to Carus (2007a, 152, n.11), Carnap (1924) "was composed just after" "Chaos" in the summer of 1922.

> Neurath says that unfortunately my book does not have the right effect ethically on those for whom it is really written, because it opposes materialism and realism much more sharply than idealism which, after all, is the worse enemy. He speaks of the world view of the new age. My book should give greater emphasis to collectivism, the 'methodological solipsism' is not to his taste. (RC 025-72-05 ASP)

Here the ultimate object of Neurath's criticism is identified: methodological solipsism! That Neurath guardedly renewed this criticism in his review shows, of course, that he did not think that Carnap's revisions had taken it on board sufficiently. To be sure, Carnap may be understood to have bracketed the issue of "collectivist" theory choice along with others that pertained to the "importance for science" of the physicalist system, which, he said, he could not discuss in "more detail" in the *Aufbau*.[32] However, there is no denying the radical nature of Neurath's challenge. To call into question the applicability of methodological solipsism to the epistemology of science is to call into question the point and relevance of the *Aufbau* as Carnap had executed it. By all means ground objectivity in structure, but seek it elsewhere, Neurath can be understood to have urged Carnap: do provide a constitutional system that exhibits a common empirical ground of all (non-formal) scientific concepts, but do not go back behind what is intersubjectively accessible. This, of course, was the motivating thought behind Neurath's understanding of physicalism and it showed little impact on Carnap in the *Aufbau*. Rather it formed, *in nuce*, the topic of the protocol sentence debate as far as it played out between Carnap and Neurath from 1929 to 1932.

2.6 Conclusion

Carnap nevertheless took Neurath's pre-publication criticism seriously for several of his points are reflected in the published book. To call Neurath's impact on Carnap's *Aufbau* "merely representational" would be to misrepresent its import. What cannot be discounted, moreover, is the possibility that it was Neurath who, in line with his stress on objectivism, also urged Carnap to say more about the possibility of a physical constitution system than he had planned originally. Stripped down to its basics in this fashion, Quine's guess that Carnap inserted the "paragraphs of disavowal" in response to his discussions with Neurath appears entirely plausible. What

---

[32] See the response to Neurath's review in Carnap to Neurath, 7 October 1928 (RC 029-16-01 ASP), discussed in Uebel (2007a, 106–108).

we must not do is to take Quine's guess to be correct in suggesting, first, that Neurath provoked a radically new insight and, second, that his criticisms of the *Aufbau* provoked a substantial change toward physicalism as we know it from 1932 on Carnap's part.

So how did Carnap manage to overcome his early attachment to the epistemological "given" in the *Aufbau* – and how did Neurath and, much later, Quine argue against it? What is notable is that all three arrived at their conclusion by different routes. Abstracting from numerous details and suppressing further qualifications, their main argumentative positions can be summarized very briefly as follows.

Having noted that a methodologically solipsist constitution system does not support the intersubjectivity of science (its statements remained unintelligible to others: all reconstructed intersubjectivity remained intersubjectivity-in-the-image), Neurath hit upon an extremely compact version of what since has become known as a private language argument: "If someone makes predictions and wants to check them himself, he must count on changes in the system of his senses, he must use clocks and rulers; in short, the person supposedly in isolation already makes use of the 'intersensual' and 'intersubjective' language" (1931, 55). If physicalistic statements about instrument readings need themselves be translated, in order to be meaningful statements, into phenomenal terms directly related to a scientist's experience, then no touchstone at all would be available by which the constancy of his language use could be established: what could such phenomenal terms refer to but my experience *now*? Neurath suggested that once on a solipsistic base there was no preventing solipsism-of-the-moment. Consequently, reconstructions of the language of unified science – which were to serve epistemological purposes after all – had to be physicalistic from the start and any appeal to the phenomenal given had to be abandoned.

Though he conceded several points to Neurath's campaign for physicalism (like the need for autopsychological statements to be physicalistically translatable), ultimately Carnap was not moved by Neurath's private language argument. Dropping the mandatory demand for methodologically solipsist reconstructions in late 1932, Carnap held on to their possibility until a logical argument convinced him of their untenability. It was in "Testability and Meaning" that he abandoned methodological solipsism when he stated explicitly that its phenomenalist language "is a purely subjective one, suitable for soliloquy only, while the intersubjective thing-language is suitable for use among different subjects" (1937, 10). For a reconstruction of the language of unified science that promised practical benefits, such a language is unsuitable. While it is possible to design so-called reduction sentences that – despite their name – non-reductively relate expressions of the thing-language to expressions in the phenomenal language, Carnap showed that was impossible to construct the thing-language on the basis of the phenomenal one or effect a "re-translation" of the former in terms of the latter (1936a, 464). (The lesson learned that disposition terms were irreducible to observational terms applied equally to the reduction of observational terms of the physical language to phenomenal terms.)

For his part, Quine rejected the "make believe" of phenomenalist reconstructions of discourse of physical objects, properties and events on yet different grounds, namely on account of what he argued to be the reductive failure of the *Aufbau* at §126. When it came to constituting the space-time world, in particular the assignment of colors to world points, it was no longer definitions of terms by previously defined lower-order ones that were provided but only desiderata. "The connective 'is at' remains an added undefined connective; the canons counsel us in its use but not its elimination" (1951f, 40). Given this failure of what he regarded as the most thorough attempt ever made to realize a phenomenalist reduction of physical object discourse, Quine drew the obvious consequence, albeit not immediately.[33] As he put it, "the trouble is that immediate experience simply will not, of itself, cohere as an autonomous domain. References to physical things are largely what hold it together." And following a reference to Neurath's simile of seafarers rebuilding their ship at sea he added: "If we improve our understanding of ordinary talk of physical things, it will not be by reducing that talk to a more familiar idiom; there is none. It will be by clarifying the connections, causal or otherwise, between ordinary talk of physical things and various further matters which in turn we grasp with help of ordinary talk of physical things" (1960, 2–3).

Here we find Quine at one with Neurath and Carnap and it is this convergence that his guess about their pre-publication discussion of the *Aufbau* anticipated. As we can see, however, their embrace of physicalism and rejection of methodological solipsism turned on considerations that did not yet figure in these pre-publication discussions. Moreover, each of Quine, Carnap and Neurath had to find their own route to their physicalist understanding of the language of science. Yet, crucially, they agreed

---

[33] For a discussion of some of the issues involved that cannot be followed up here, see Verhaegh (2018, ch. 2).

that the doctrine of methodological solipsism failed to deliver the reduction that its success would have required.

Archival Sources

ASP = Carnap Nachlass, Archives of Scientific Philosophy, Special Collections, Hillman Library, University of Pittsburgh, Pittsburgh

WKA = Neurath and Schlick Nachlass, Wiener Kreis Archief, Noord-Hollands Archief, Haarlem, The Netherlands

CHAPTER 3

Frameworks, Paradigms, and Conceptual Schemes
Blurring the Boundaries between Realism and Anti-Realism

Sean Morris

On the standard story, the demise of logical positivism is chiefly attributed to W. V. Quine's 1951 "Two Dogmas of Empiricism" and Thomas Kuhn's 1962 *The Structure of Scientific Revolutions*. Quine, by way of rejecting the analytic/synthetic distinction, knowingly "blurr[ed] the supposed boundary between speculative metaphysics and natural science" (1951f, 20) and has been taken to let back in the sort of metaphysics that logical positivism had sought to vanquish. And from another direction, Kuhn, by way of his historical and sociological approach to the philosophy of science, has been taken to show that the purely formal methods of logical empiricism and a sharp distinction between theory and observation are untenable as a philosophy of science that accounts for the actual practice of science. In this essay, I challenge this view of Quine and Kuhn on the one side and logical positivism – by way of Rudolf Carnap – on the other.[1] While I do not mean to discount or dismiss important differences among these figures, I will argue that they share a core commitment to reconceiving of ontology that undermines traditional realist and anti-realist approaches to it. All propose an account of reality that answers only to science itself, and in doing so, blurs the boundaries between realism and anti-realism.

The structure of this essay is as follows. I begin with a very brief account of the realism that Carnap, Quine, and Kuhn reject, and then lay out the basic details of Carnap's approach to ontology. This provides

---

[1] Since the 1990s, a number of scholars have increasingly read Kuhn as sharing key features of Carnap's philosophy, particularly the notion of a framework, which for Kuhn becomes a paradigm. For this reading of Kuhn, see, for example, Reisch (1991); Friedman (2003) and (2012); and Richardson (2007). Quine at one point declared his aim in rejecting the analytic/synthetic distinction as "being more Carnapian than Carnap" (Quine 1994b, 154). For a recent account of how much Quine shares with Carnap, see Verhaegh (2018), especially ch. 3.

the background against which both Quine's and Kuhn's views can be profitably considered.

3.1

Before turning directly to Carnap, let me briefly characterize the sort of realism that Carnap, Quine, and Kuhn set their views against. The view is that there must be a reality that grounds our scientific theories, one that is prior to or stands outside science. Following Putnam, I will call this metaphysical realism.[2] In an ideal case, a scientific theory is true when it describes reality. In nonideal cases, some theories still count as better than others because they describe reality more accurately than others, even if they do not do so perfectly. The central problem with this view is that it is unclear how we could know that a scientific theory has accurately, or *more* accurately, described, a reality that stands prior to science itself. How could we ever have access to such reality?[3] This sense of realism may seem naïve given the far more sophisticated versions of scientific realism that began to emerge in the early 1960s.[4] I will not argue directly for it in this essay but would claim that these more sophisticated views could emerge only against the backdrop of Carnap, Quine, and Kuhn. Let me turn now directly to Carnap.

Throughout his career, Carnap never sought to settle disputes between realists and anti-realists.[5] Rather, he saw such disputes as responsible for philosophy's lack of progress – its "empty wrangling" – in contrast to the clear progress found in the natural sciences (Carnap 1935, 81). In his 1928 *Aufbau*, Carnap sought a metaphysically neutral position between

---

[2] In adopting Putnam's terminology here, I am not claiming that the sort of realism I describe here is exactly what he had in mind. Kuhn specifically contrasts his view with Putnam's notion of metaphysical realism (Kuhn 1990, 101), and I think this accurately characterizes the realism that Carnap and Quine oppose as well. I will not adopt Putnam's contrasting notion of internal realism to describe the views of Carnap, Quine, and Kuhn, though in a broad sense, this is a fair characterization of their respective versions of a realism internal to science. Carnap thinks his view is only misleadingly described as realism; Kuhn never adopts realism as a description of his view; and while Quine does, we must keep in mind that his realism takes a very specific naturalized form. Putnam introduces the distinction in his (1976).

[3] Bertrand Russell in certain periods of his work holds something like this sort of realism where the relation of acquaintance is meant to give direct access to reality. For more on this view of Russell, see Hylton (2004), 115–116. Carnap, in the *Aufbau*, cites Russell as an example of a metaphysical realist (Carnap 1928/1967, sec. 175–176).

[4] See, for example, Grover Maxwell (1962).

[5] Although Carnap has often been taken by contemporary anti-realists as a precursor to their views; see, for example, Bas van Fraassen (1980). We will later see that something similar has been done with regard to Kuhn on the anti-realist side and with regard to Quine on the realist side. Throughout, I use the more general contemporary term 'anti-realists', though Carnap most often characterizes the debate as specifically between realists and idealists. In cases where it matters, I will be more specific about the form of anti-realism under consideration.

disputants, a view he later described as his earlier less radical critique of traditional metaphysics (Carnap 1963a, 18–19).[6] His more radical rejection of traditional metaphysics began to emerge in the early 1930s, reaching its mature phase with his adoption of the principle of tolerance in his 1934 *The Logical Syntax of Language*.[7] Here, metaphysical views were not rejected in the flatfooted way of the early 1930s. Rather, tolerance allowed philosophers to adopt any language they might like and offer pragmatic reasons as to why one language rather than another might be preferred. According to Carnap: "*In logic, there are no morals.* Everyone is at liberty to build up his own logic, i.e. his own form of language, as he wishes. All that is required of him is that, if he wishes to discuss it, he must state his methods clearly, and give syntactical rules instead of philosophical arguments" (Carnap 1934/1937, 52; Carnap's italics). In his 1950 "Empiricism, Semantics, and Ontology," Carnap presented his most developed version of this view, with a particular emphasis on ontological questions.[8] I will focus on this essay in the remainder of this section.

As in his earlier discussions of realism, Carnap does not aim here to settle disputes between realists and anti-realists,[9] but rather seeks to show that both positions lie outside of a properly scientific philosophy. More specifically, he aims to show how empiricists can adopt the abstract objects of mathematics and semantics without falling into idle metaphysical disputes over their ontological status, though he will offer similar considerations for entities of any sort, including physical objects. Carnap begins by observing that to speak about a new kind of entity,[10] we must introduce a system for speaking about it with appropriate new rules. He refers to such a system as a linguistic framework. Correspondingly, two kinds of existence questions arise. One kind concerns the existence of entities within the framework, which Carnap calls internal questions. The other kind concerns the existence of the system of entities as a whole, which he calls external questions.

---

[6] For a detailed account of Carnap's rejection of metaphysics in the *Aufbau*, see Friedman (2007).
[7] For an account of the development of Carnap's anti-metaphysical position in the early 1930s, see Creath (2012b); for the changes that tolerance brought to it, see Creath (2009). I follow Creath's account in my account of the anti-metaphysical position that Carnap developed in the early years of the 1930s.
[8] Carnap included this article (with some changes) as an appendix in the 1956 second edition of *Meaning and Necessity*. I refer to this later version.
[9] Carnap tends to oppose realism to idealism, but I have chosen to use the more general term "anti-realism" for ease of exposition throughout. In cases where the specific version of anti-realism matters, I refer to the more specific case.
[10] These may be entities already found in language but now subjected to explicit rules for their use – for example, the physical objects of everyday language (Carnap 1950/1956, 207).

We formulate internal questions and possible answers to them with the help of the new forms of expression. Answers are then found by purely logical means or by empirical methods, depending on whether the framework is logical or factual. To illustrate his point, Carnap turns to what he considers the simplest kind of entities, the spatio-temporally ordered system of observable things and events, which he calls the world of things. Once we adopt a language for them, we can answer internal questions about them by empirical methods, specifically by way of observations evaluated according to rules of confirmation and disconfirmation. Reality, in this sense, is just an ordinary empirical, scientific, nonmetaphysical concept.

External questions, on Carnap's account, are quite different from internal questions, as they concern the reality of the thing world itself and are raised only by philosophers.[11] It is these sorts of question that realists and anti-realists dispute endlessly. According to Carnap, they are framed in the wrong way from the start. To be real in the scientific sense is, as noted, to be an element of a linguistic framework, so the system itself cannot be said to be real (or unreal) (Carnap 1950/1956, 206–207). Carnap then offers another way of understanding such disputes, remarking that those who raise these questions may not mean them as theoretical questions about the reality or unreality of the thing world but as practical questions about whether to accept the linguistic framework of things. If the disputes are understood in this way, Carnap has no objection to saying that someone has accepted the world of things. He is careful to note, however, that this does not indicate belief in the reality of the thing world. Since the question of whether to accept the thing language is not of a theoretical nature, there is no such belief or assertion about the reality of thing world itself. The acceptance of the thing language cannot be formulated within the thing language (Carnap 1950/1956, 207–208). Still, while the acceptance of the thing language is not itself of a theoretical nature, it will be influenced by theoretical knowledge:

> The purposes for which the language is intended to be used, for instance, the purpose of communicating factual knowledge, will determine which factors are relevant for the decision. The efficiency, fruitfulness, and simplicity of the use of the thing language may be among the decisive factors. And the questions concerning these qualities are indeed of a theoretical nature. But these questions cannot be identified with the question of realism. They are

---

[11] Carnap has two senses of external questions: the one described here and also, as we will see, one that concerns the practical adoption of a linguistic framework. On this point, see Eklund (2013), 237, and Verhaegh (2018), 42–43.

not yes-no questions but questions of degree. The thing language in the customary form works indeed with a high degree of efficiency for most purposes of everyday life. This is a matter of fact, based upon the content of our experiences. However, it would be wrong to describe the situation by saying: "The fact of the efficiency of the thing language is confirming evidence for the reality of the thing world"; we should rather say instead: "This fact makes it advisable to accept the thing language." (Carnap 1950/1956, 208)

I quote this passage at length as these issues over the boundaries between practical and theoretical decisions, which Carnap sees as clearly distinguished and of a different kind in principle, are crucial to understanding the similarities and differences between Quine and Carnap, and also, in a way, Kuhn.

Carnap concludes by summarizing his procedure for introducing a new kind of entity, as follows. We represent the acceptance of entities by introducing a framework of forms of expression, along with a set of rules for using these expressions. New names may be introduced into a framework, but this is not necessary. The two crucial steps are (1) to introduce a general term, that is, a predicate of higher level, for the new kind of entity, one that says of any particular entity that it belongs to that kind; and (2) to introduce variables of the new type, for which the new entities are available as values. Here, Carnap claims agreement with Quine, praising him for being "the first to recognize the importance of the introduction of variables as indicating the acceptance of entities" (Carnap 1950/1956, 214, fn. 3).[12] Given that these variables can be used to make general statements about new entities, internal questions can be formulated and answered about them (Carnap 1950/1956, 213–214).

Carnap then carefully distinguishes the external questions that philosophers have asked concerning the existence or reality of the total system of new entities, which they take to require an answer prior to the introduction of new linguistic forms. For these philosophers, such introductions will be legitimate only if some ontological insight can provide an affirmative answer to the question of the reality of the entities in question. In contrast, Carnap urges his view that introducing new ways of speaking requires no theoretical justification because they make no assertions regarding what exists fundamentally. To speak of the acceptance of new entities indicates only that we accept the new linguistic framework. It makes no claims about the reality of the entities concerned, but intimates only that its use satisfies certain practical aims. To treat such a practical question otherwise is to pose a pseudo-question (Carnap 1950/1956, 214).

---

[12] Whether Carnap really does agree with Quine is open to dispute; see Roberta Ballarin's contribution to this volume.

Since the acceptance of a linguistic framework does not imply any metaphysical doctrine about the reality of the entities under consideration, Carnap holds, as a final consequence, that empiricists need have no worries about accepting or rejecting abstract entities (Carnap 1950/1956, 214–215). The acceptance or rejection of linguistic forms of any sort, he concludes,

> in any branch of science, will finally be decided by their efficiency as instruments, the ratio of the results achieved to the amount and complexity of the efforts required. To decree dogmatic prohibitions of certain linguistic forms instead of testing them by their success or failure in practical use, is worse than futile; it is positively harmful because it may obstruct scientific progress. The history of science shows examples of such prohibitions based on prejudices deriving from religious, mythological, metaphysical, or other irrational sources, which slowed up the developments for shorter or longer periods of time. Let us learn from the lessons of history. Let us grant to those who work in any special field of investigation the freedom to use any form of expression which seems useful to them; the work in the field will sooner or later lead to the elimination of those forms which have no useful function. *Let us be cautious in making assertions and critical in examining them, but tolerant in permitting linguistic forms.* (Carnap 1950/1956, 221; Carnap's italics)

Carnap, as I read him, shifts our understanding of questions of realism versus anti-realism in ontology by taking such questions, asked as theoretical questions, to be pseudo-questions. In this way, there is no dispute to be settled. To the extent that such questions have any place at all, it is as external questions, practical questions of whether or not to accept a particular linguistic framework. From his perspective, the internal sense of reality provides all the reality that science needs. Let me now consider Quine.

3.2

Quine, unlike Carnap, is a self-avowed realist.[13] Although an explicit commitment to realism does not appear in his work until much later, already at the opening of his 1951 "Two Dogmas of Empiricism," Quine announced, as a consequence of rejecting the analytic/synthetic distinction, the "blurring of the supposed boundary between speculative metaphysics and natural science" (Quine 1951f, 20).[14] In particular, this puts "ontological

---

[13] For an example, see Quine (1981d), 21.
[14] Whether it is really the analytic/synthetic distinction that lies at the bottom of the distinction between matters of fact and matters of choosing a linguistic framework is a question of some dispute. Quine and Carnap themselves seem to see it this way, but see Peter Hylton's contribution to this volume, where he argues that it is really the principle of tolerance that is at issue.

questions ... on a par with questions of natural science" (Quine 1951f, 45).[15] This is not yet Quine's realism, but I will argue that it provides the basis for it. Once he rejects the idea of an in-principle distinction between ontological questions and questions of natural science, Quine can recognize only one sense of reality (Quine 1957, 229), that given by science itself. It is here that his realism emerges full-blown, a realism that in no way stands outside of natural science, somehow serving to ground it. At its basis is his naturalism, that there is no first philosophy prior to science. In this section, I will first discuss Quine's 1948 "On What There Is," his first comprehensive statement on ontology, and will then turn to consider his later claims of realism. My aim is to bring out the sense in which realism is internal rather than prior to science.

Quine discusses the sense in which ontological questions are on par with questions of natural science rather briefly in "Two Dogmas" but goes into more detail in the 1948 "On What There Is." In the first half of this paper, he develops his views on ontological commitment, concluding with his now familiar criterion: "To be is to be the value of a variable" (Quine 1948b, 13 and 15).[16] As we saw, with regard to ontological commitment, Carnap claims agreement with Quine. Following this discussion, however, Quine turns to broader scale ontological concerns. Here, differences with Carnap emerge. This comes out in Quine's discussion of questions of competing ontologies.

Besides agreeing over the criterion for ontological commitment, Quine would also seem to agree with Carnap on competing ontologies, remarking that "the obvious counsel is tolerance and an experimental spirit" (Quine 1948b, 19). Again, Carnap himself notes this: "With respect to the basic attitude to take in choosing a language form (an 'ontology' in Quine's terminology, which seems to me misleading), there appears now to be agreement between us" (Carnap 1950/1956, 215, fn. 5). Quine is indeed tolerant but, in contrast with Carnap, not in a way that he takes to obviate ontological commitments.

Quine offers an early version of his view that ontological questions are not to be distinguished from scientific questions and that the acceptance of an ontology is on a par with the acceptance of a scientific theory:

---

[15] In the version of "Two Dogmas" reprinted in *From a Logical Point of View*, Quine adds a footnote to Emile Meyerson in support of this view. Kuhn also cites Meyerson as a major influence on his work generally (Kuhn 1962/1970, vii–viii).

[16] This is not the first place such views appear. Quine had put forth the criterion as early as 1939 (Quine 1939, 708).

> Our acceptance of an ontology is, I think, similar in principle to our acceptance of a scientific theory, say a system of physics: we adopt, at least insofar as we are reasonable, the simplest conceptual scheme into which the disordered fragments of raw experience can be fitted and arranged. Our ontology is determined once we have fixed upon the over-all conceptual scheme which is to accommodate science in the broadest sense; and the considerations which determine a reasonable construction of any part of that conceptual scheme, for example, the biological or the physical part, are not different in kind from the considerations which determine a reasonable construction of the whole. To whatever extent the adoption of any system of scientific theory may be said to be a matter of language, the same – but not more – may be said of the adoption of an ontology. (Quine 1948b, 16–17)

In contrast to Carnap, Quine finds no distinction to draw between adopting an ontology and adopting a scientific theory. Adopting an ontology is not a matter of adopting on pragmatic grounds a linguistic framework within which scientific theories can be formulated. For Quine, the pragmatic grounds for adopting an ontology are just as much in play when adopting a specific scientific theory. To the extent that either is a matter of language, so is the other. And without a fixed boundary between the linguistic and the factual, our ontology is just as factual as the rest of science. Contrary to Carnap, appealing to language does not render ontological claims philosophically, or scientifically, inert.

To my knowledge, Quine himself does not describe himself in print as a realist until the mid-1970s, in the paper "The Pragmatists' Place in Empiricism," though by the 1960 *Word and Object*, he took realism to be an accurate description of his philosophy of science.[17] Still, though his views in 1948 were still not fully settled – in particular, Quine continued to be open to phenomenalist approaches to epistemology[18] – the view put forward here opens the door to the sort of robust realism that Quine would later adhere to.[19] The key point to

---

[17] "The Pragmatists' Place in Empiricism" was not published until 1981. J. C. C. Smart, in his contribution to the 1969 volume on Quine's philosophy, *Words and Objections*, describes Quine's philosophy of science as distinctly realist as early as the 1960 *Word and Object*, though noting earlier phenomenalist and instrumentalist tendencies. To the realist description, Quine responds affirmatively. Although not stated publicly, Quine described himself as a realist as early as 1944 (Verhaegh 2018, 93 and 173).

[18] Although Quine already believed the reduction of science as a whole to sense data would not work (Quine 1948b, 17–18). See Hylton (2007), 85–87, and Verhaegh (2018), 28–29, as to why Quine ultimately rejected phenomenalism. See Verhaegh (2018), 26–27; and for a detailed account of Quine's evolving views in this period, see Verhaegh (2018), ch. 5, especially 94–95 for the development of his views through 1948.

[19] Quine describes his view this way, though, as we will see, his realism may not be as robust as metaphysical realists would want it to be. For Quine, though, there is no more robust standpoint than that of science itself.

draw from this period is that whatever ontology we ultimately commit ourselves to, our reasons for doing so should come from scientific methodology itself.[20] We adopt the conceptual scheme that best organizes our experience, appealing to such pragmatic considerations as simplicity when we do so. To do otherwise would be "a scientific error" (Quine 1951f, 44–45).

To return to the 1975 paper just mentioned, here Quine describes the pragmatists' view of science "as a conceptual shorthand for organizing experience," and then observes, despite his own avowal of naturalism, that he "seem[s] to be drawn into the same position. Is there no difference?" (Quine 1975, 33).[21] His realist ontology is precisely where Quine finds the difference:

> For naturalistic philosophers such as I … physical objects are real, right down to the most hypothetical particles, though this recognition of them is subject, like all science, to correction. I can hold this ontological line of naive and unregenerate realism, and at the same time I can hail man as largely the author rather discoverer of truth. I can hold both lines because the scientific truth about physical objects is still the *truth*, for all man's authorship. In my naturalism, I recognize no higher truth than that which science provides or seeks. (Quine 1975, 33–34; Quine's italics)

Quine goes on to explain that scientists are indeed creative in their work, positing physical objects, and that they perhaps could have produced a different system that would have fit the data just as well. This is, though, not to somehow rise above science. Rather, Quine explains,

> [T]o say all this is to affirm truths still within science, about science. These truths illuminate the methodology of our science but do not falsify or supersede our science. We make do with what we have and improve it when we see how. We are always talking within our going system when we attribute truth; we cannot talk otherwise. Our system changes, yes. When it does, we do not say that truth changes with it; we say that we had wrongly supposed something true and have learned better. Fallibilism is the watchword, not relativism. Fallibilism and naturalism. (Quine 1975, 34)

By recognizing no clear distinction between a change of theory formulated within a linguistic framework and a change in the framework itself, Quine does not hold, as Carnap does, that ontological claims are merely pragmatic,

---

[20] Quine himself had not yet fully appreciated this point, remarking that it was in 1950s that he became "more consciously and explicitly naturalistic" (Quine 1991, 398). Verhaegh emphasizes this remark (Verhaegh 2018, 79).

[21] Verhaegh (2018) draws on this paper as well in his account of Quine's realism (61–62). I am much influenced by his interpretation.

conventional choices of language.²² Indeed, from Quine's perspective, Carnap's remark that though accepting a particular linguistic framework is not of a theoretical nature, such a decision will be influenced by theoretical knowledge (Carnap 1950/1956, 208) opens the way to Quine's view. For him, this remark indicates that there is no clear place to draw such a boundary between the pragmatic and the theoretical.²³ As we have seen, Quine agrees that accepting an ontology may be a matter of choosing a convenient language form for science, but as this, for him, *is* a matter of scientific theorizing, such decisions generate genuine ontological claims. The grounds may be pragmatic, but such grounds are just as much a part of science as observation. Our posits, made as part of the theory-building process, are to be taken seriously from the standpoint of that theory. In taking the theory to be true, we accept as real the objects that that theory says there are. For Quine, "To call a posit a posit is not to patronize it" (Quine 1960, 22).

There is still, Quine admits, fallibilism about our theory, but not, he insists, relativism about it. This is not incompatible with his realism but integral to it. As he later explains: "[T]here is an absolutism, a robust realism, that is part and parcel of my naturalism. Science itself, in a broad sense, and not ulterior philosophy, is where judgment is properly passed, however fallibly, on questions of truth and reality. Whatever is affirmed there, on the best available evidence, is affirmed as absolutely true" (Quine 1984, 321). Science itself is fallible, and to put forth a philosophy that tries to make it otherwise offends against elementary fact (and Quine's naturalism).

Still, despite his differences with Carnap, we should not lose sight of what Quine shares with Carnap here. Realism on Quine's view is a part of natural science, not something that stands outside of science and tries to ground it extra-scientifically. As Quine sees it, realism is not a view subject to some scientifically prior philosophy: it is itself part of science. If this is still a metaphysical view, it should be recognized that it is metaphysics naturalized.²⁴ Part of Quine's rejection of extra-scientific realism is the inclusion of an anti-realist – specifically, an instrumentalist – element in his view, that of positing the entities of science so as to better cope with sensory experience. But here we also get his particular form of realism: that from the perspective of the theory, these entities are as real as any science can sustain. We saw above Quine appealing to his robust realism to distinguish his view from the

---

[22] For more on this point, see Hylton's contribution to this volume.

[23] Quine noted on the back of a 1949 letter from Carnap concerning ontology, "Say frameworkhood is a matter of degree, & reconciliation ensues" (Quine and Carnap 1990, 417).

[24] On Quine's view as metaphysics naturalized, see Hylton (2014a). For the resurgence of metaphysics after Quine, see Hylton (2014); Kemp (2014) and (this volume); and Rosen (2014).

instrumentalism of the pragmatists. Theirs is an instrumentalism that stands outside of science.[25] His realism, in contrast, does not. It is a realism fully internal to science.[26] Despite his own avowals of a robust realism, we might do better, in a sense, to say that Quine is neither a realist nor anti-realist, at least not in any traditional philosophical sense of these terms.[27]

Let me turn now to consider where Thomas Kuhn stands on issues of realism and anti-realism and also where he stands in relation to Carnap and Quine.

3.3

Kuhn, unlike Carnap and Quine, is less directly concerned with issues of ontology, but much of *The Structure of Scientific Revolutions* has implications for ontology. Indeed, Kuhn has been taken as a central opponent to any kind of realism in the philosophy of science. Mostly he is held to favor some form of anti-realism, even postmodernism.[28] In support of such readings of Kuhn, he is said to believe that scientists from different paradigms work in different worlds and there is no progress in science.[29] At the very least, such views take him to rule out any sort of objective reality for science to be measured against. But he is often portrayed as holding something far stronger: the complete relativity of all scientific theories. While Kuhn may sometimes seem to believe this, he never construed his aims as the rejection of realism. Instead, in a retrospective comment he describes his purpose as follows:

---

[25] Here, I am especially indebted to Verhaegh (2018).

[26] In his "Identity, Ostension, and Hypostasis," Quine writes, "[W]e cannot detach ourselves from our [conceptual scheme] and compare it objectively with an unconceptualized reality. Hence it is meaningless, I suggest, to inquire into the absolute correctness of a conceptual scheme as a mirror of reality. Our standard for appraising basic changes of conceptual scheme must be, not a realistic standard of correspondence to reality, but a pragmatic standard" (1950/1980, 79). I take it that my reading of Quine's realism reconciles his view with this passage. His denial of a realist standard is of the traditional sort sketched at the outset of this paper. Quine's realism is wholly internal to science, as is the pragmatism inherent to it.

[27] Burton Dreben made this point with regard to Quine when observing that Putnam reads Quine at times as a realist and then at others as a positivist. Dreben concludes, "We see Quine as both – hence as neither" (Dreben 1992, 301). I take it that the sense in which Quine is both is in holding to aspects of realism and anti-realism that are wholly internal to science. He is neither in denying any form of realism or anti-realism that stands prior to science.

[28] See, for example, Devitt (2008). Also see Edmister and O'Shea (1994), 44. In the introduction to their interview, the interviewers describe Quine's importance for Kuhn as follows: "Further, Quine argued that our knowledge forms a 'web of belief.' For Quine, no piece of knowledge was sacred or beyond potential revision. His work in the 1950s partially anticipates and lays the groundwork for Thomas Kuhn's *The Structure of Scientific Revolutions* (1962), a work at the very basis of the postmodern canon. The idea expressed therein – that scientific theories are simply descriptions of observed phenomena rather than deep descriptions of 'reality' – is a clear extension of Quine's web of beliefs."

[29] Kuhn does make similar, though not identical, claims. For an account of the Kuhn of legend compared to the actual Kuhn, see Sharrock and Read (2002), "Introduction."

My goal is double. On the one hand, I aim to justify claims that science is cognitive, that its product is knowledge of nature, and that the criteria it used in evaluating beliefs are in that sense epistemic. But on the other, I aim to deny all meaning to claims that successive scientific beliefs become more and more probable or better and better approximations to the truth and simultaneously to suggest that the subject of truth claims cannot be a relation between beliefs and a putatively mind-independent or "external" world. (Kuhn 1993, 330)

Although Kuhn wrote these remarks some nearly thirty years after *The Structure of Scientific Revolutions*, I think they accurately characterize his earlier work. In what follows, I will take up the two aspects of Kuhn's double goal to bring out the sense in which he not so much denies realism but, like Carnap and Quine, aims to shift our understanding of the categories of realism and anti-realism so as to bring reality firmly within the bounds of science itself.[30]

First, let me consider the way in which Kuhn sees his view of science as being cognitive, that is, as yielding genuine knowledge about the natural world. It is tempting to view Kuhn as a relativist because of passages where he appears to deny a single world which science describes. For example, Ch. X, "Revolutions as Changes of World View," opens with the claim that

> Examining the record of past research from the vantage of contemporary historiography, the historian of science may be tempted to exclaim that when paradigms change, the world itself changes with them. Led by a new paradigm, scientists adopt new instruments and look in new places. Even more important, during revolutions scientists see new and different things when looking with familiar instruments in places they have looked before. ... [P]aradigm changes do cause scientists to see the world of their research-engagement differently. In so far as their only recourse to that world is through what they see and do, we may want to say that after a revolution scientists are responding to a different world. (Kuhn 1962/1970, 111)

One certainly could take this to indicate a commitment to some form of anti-realism in that it claims that the world is not a fixed reality independent of human cognition. In a sense, this is Kuhn's view, but his considered view is more complex and, much like Quine's, incorporates aspects of both realism and anti-realism.[31]

In response to his realist critics' charges, Kuhn admits that he had not always been clear on the sense in which he holds that with a new

---

[30] In reading Kuhn this way, I have been influenced by Sharrock and Read (2002).
[31] Some readers have stated that Kuhn is in fact a realist. Since Kuhn does not apply this label to himself, I do not want to go this far. But I am sympathetic to viewing Kuhn as a realist of a sort. See, for example, Sharrock and Read (2002), especially ch. 5; and Massimi (2015).

paradigm, the world changes.[32] In his "Postscript," added to the 1970 edition of *The Structure of Scientific Revolutions*, he explains his view by offering a distinction between stimuli and sensations.[33] He states that, solipsism aside, two people standing in the same place and looking in the same direction will receive roughly the same stimuli, these being of the shared world.[34] It is not the stimuli, however, that people immediately perceive, as knowledge of them is highly abstract and theoretical. What people have instead, are sensations, and there is no reason to suppose that these are the same for different people since there is much neural processing that takes place between the receipt of a stimulus and the awareness of sensations. Included in the little that we know of this process is the fact that different stimuli can produce the same sensation; that the same stimuli can produce different sensations; and, most importantly for my point, that the path from stimuli to sensations is partially conditioned by education (Kuhn 1962/1970, 192–193).

In support of this point, Kuhn observes that people raised in different societies behave at least sometimes as if they see different things, and remarks that – the temptation to identify stimuli one-to-one with sensations aside – "we might recognize that they actually do so," that is, see different things (Kuhn 1962/1970, 193). He elaborates:

> Notice now that two groups, the member of which have systematically different sensations on receipt of the same stimuli, do *in some sense* live in different worlds. We posit the existence of stimuli to explain our perceptions of the world, and we posit their immutability to avoid both individual and social solipsism. About neither posit have I the slightest reservation. But our world is populated in the first instance not by stimuli but by the objects of our sensations, and these need not be the same, individual to individual or group to group. (Kuhn 1962/1970, 193)

---

[32] I will keep my discussion to *The Structure of Scientific Revolutions* and the works immediately following it, where Kuhn tried to clarify his views. He continued to develop them on this point for the rest of his career.

[33] I should note that this was not Kuhn's settled view on the issue of a shared reality. In further writings he would try to account for his view with a more linguistic approach and then with a more evolutionary approach, this latter approach, as we will see, having its roots in *The Structure of Scientific Revolutions*. For more on how Kuhn's views continued to evolve after the period under consideration here, see Sharrock and Read (2002), ch. 5.

[34] Kuhn does not say what the stimuli are. I think that this is on purpose since to describe them as, say, physical objects would be to already to interpret them according to a paradigm. My guess is that the stimuli, while posited as the shared world of the observers, ultimately should be subject, in part, to a paradigm for their specific characteristics. This seems to be what Kuhn proposes in stressing the "mutual plasticity" of the phenomenal and real world in his later work; see his (1990), 102.

Here, by way of positing stimuli to explain our perceptions, Kuhn makes space for a shared reality from within his view of scientists from different paradigms inhabiting different worlds. The education, training, and practices that come with different paradigms shape how scientists perceive the stimuli that impinge on them, but this does not mean that there is no shared reality. To echo Quine's point from the previous section, "To call a posit a posit is not to patronize it" (Quine 1960, 22). The positing of stimuli, for Kuhn, provides the shared reality that our sensations are beholden to. Members of the same scientific community will share such things as education, training, and practices, and because of this we may expect their sensations to be the same (Kuhn 1962/1970, 193). But scientists from other paradigms, who have different education, training, and practices, will, we can be pretty sure, experience the world differently. In this sense, they live in a different world. In both cases the stimuli are what the prior education, training, and practice associated with a paradigm apply to and are what generate sensations.[35] If this is what Kuhn indeed originally intended in *The Structure of Scientific Revolutions*, he has to be credited with believing in a fixed reality independent of human cognition. Kuhn's view here does not stand outside of science itself. The positing of stimuli to explain our sensations is itself a part of scientific methodology. And so Kuhn's claim that scientists of different paradigms inhabit different worlds in no way detracts from his account of science. This follows unless we think that philosophy should provide the kind of reality that science has no truck with – a reality that stands apart from all known sources of human cognition. On this reading of Kuhn's thinking, he aims to provide us with a sense of reality that stands firmly within science.

Let me now turn to Kuhn's second aim, his denial that progress in science is constituted by successive scientific theories better and better approximating an external reality. Again, realist critics have taken Kuhn's view here to favor a form of relativism incompatible with realism properly understood. As we will see, Kuhn admits that this is perhaps true with regard to relativism. But with regard to realism, Kuhn again rejects only a form of realism that goes beyond what he thinks science can deliver. As with the previous criticism, Kuhn responds by indicating that traditional realism is misguided in its aims and that his own view does not deny that

---

[35] I take it that this is what Sharrock and Read are suggesting when they write: "[T]here must also be something in the visual field which is not a product of prior experience, for prior experience must be applied to something to generate an observation. And it is, of course, to an input from nature itself that prior experience must be applied" (Sharrock and Read 2002, 178).

science progresses. More specifically, he asks us to reconceive what such progress looks like. He explains his opponents' charge of relativism in terms of incommensurability – specifically, the unavailability of translations between different scientific theories. "The proponents of different theories are," he writes, "like the members of different language-culture communities. Recognizing the parallelism suggests that in some sense both groups may be right. Applied to culture and its development that position is relativistic" (Kuhn 1962/1970, 205). So, with regard to culture, Kuhn agrees that there is relativism. He goes on to explain, however, that applying this view to science, which is his concern, turns out otherwise, and at the very least "is very far from *mere* relativism" (Kuhn 1962/1970, 205; Kuhn's italics). Here, his critics have failed to recognize the significance that he places on the important role of puzzle-solving within the developed sciences and his claim that the practitioners of these sciences are "fundamentally puzzle-solvers" (Kuhn 1962/1970, 205). Kuhn explains:

> Though the values that they deploy at times of theory-choice derive from other aspects of their work as well, the demonstrated ability to set up and to solve puzzles presented by nature is, in case of value conflict, the dominant criterion for most members of a scientific group. Like any other value, puzzle-solving ability proves equivocal in applications. Two men who share it may nevertheless differ in the judgments they draw from its use. But the behavior of a community which makes it preeminent will be very different from that of one which does not. (Kuhn 1962/1970, 205)

For Kuhn, the preeminence of puzzle-solving is a key indicator of the presence of a scientific community.[36]

With puzzle-solving accorded its rightful place, Kuhn says we can imagine a tree that traces the evolution of the modern scientific specialties from their beginnings in primitive natural philosophy and the crafts. A line drawn up the tree from the trunk through to some branch tip, never doubling back, would trace successive theories related by descent. It should then be fairly easy, Kuhn thinks, to draw up some list of criteria – for example, accuracy of prediction, balance between esoteric and everyday subject matter, and the number of problems solved – that would allow an observer to distinguish earlier from more recent theories. Such a completed list would, Kuhn writes, show that "scientific development is, like biological, a unidirectional and

---

[36] For more on the role of puzzle-solving as a mark of scientific communities within the original text of *The Structure of Scientific Revolutions*, see Kuhn (1962/1970), 204–205. Here, Kuhn gives a list of additional features that characterize a scientific community, but, in one way or another, they are all connected back to puzzle-solving.

irreversible process. Later scientific theories are better than earlier ones for solving puzzles in the often quite different environments to which they are applied." This, he concludes, "is not a relativist's position, and it displays the sense in which I am a convinced believer in scientific progress" (Kuhn 1962/1970, 205–206).[37] On Kuhn's view then, contrary to what many realists have claimed, scientific theories do not progress by moving closer and closer to the truth about some single fixed reality, just as biological organisms do not progress by moving closer and closer to some prior, determinate nature that they are supposed to have. Instead, in both cases, progress is a matter of improved coping with problems encountered. In the case of the biological organism, such evolution increases the organism's ability to cope with the environment. In the case of the scientific theory, new theories allow for better solving of the problems within the discipline. In both cases, the prior stage in the evolution dies out.[38] Although Kuhn does not hold to a view that science is progressing toward some prior, fixed goal, any more than a biological organism is, he does hold to a very definite nonrelativistic notion of progress.

Kuhn acknowledges, though, that his account of progress lacks an element that most philosophers of science take to be essential to the notion of scientific progress – that in addition to being better at discovering and solving puzzles, a successor theory moves us closer to reality, that "it is somehow a better representation of what nature is really like." Here, the thought is ontological, that the successor theory is a better match "between the entities with which the theory populates nature and what is 'really there'" (Kuhn 1962/1970, 206). Then, echoing both Carnap and Quine, Kuhn concludes, "There is, I think, no theory-independent way to reconstruct phrases like 'really there'; the notion of a match between the ontology of a theory and its 'real' counterpart in nature now seems to me illusive in principle" (Kuhn 1962/1970, 206). This he claims is true not merely on philosophical grounds but also on historical grounds. He writes: "I do not doubt, for example, that Newton's mechanics improves on Aristotle's and that Einstein's improves on Newton's as instruments for puzzle-solving. But I can see in their succession no coherent direction of ontological development." In some significant ways, he maintains, Einstein's general theory

---

[37] In his (1970), Kuhn offers a similar defense of his view. Here, he again says that there is a sense in which he may be a relativist, but in the more essential sense – that "[o]ne scientific theory is not as good as another for doing what scientists normally do" – he is not a relativist (Kuhn 1970, 264).

[38] With regard to scientific theories, Kuhn quotes Max Planck on this point in the original text of *The Structure of Scientific Revolutions*: "A new scientific truth does not triumph by convincing its opponents and making them see the light, but rather because its opponents eventually die, and a new generation grows up that is familiar with it" (Planck 1950, 33–34), quoted in Kuhn (1962/1970), 151.

of relativity is ontologically closer to Aristotle's theory than either theory is to Newton's (Kuhn 1962/1970, 206; see also Kuhn 1970, 265). Kuhn does not find his position relativistic, but even if it is, he concludes, he sees no loss in how well it accounts for the nature and development of the sciences (Kuhn 1962/1970, 206–207).

Throughout this section, we have seen that Kuhn is able to do what his realist critics charge that he cannot do – make provision for reality. Yet he does not deny certain anti-realist aspects of his view, for example, his positing of stimulations or the relativism inherent in his notion of a paradigm. As in the case of Carnap and Quine, by mixing elements of both realism and anti-realism, Kuhn offers an account of reality that he sees as squaring with science itself rather than prior philosophy:

> Can a world that alters with time and from one community to the next correspond to what is generally referred to as the "real world"? I do not see how its right to that title can be denied. It provides the environment, the stage, for all individual and social life. On such life it places rigid constraints; continued existence depends on adaptation to them; and in the modern world scientific activity has become a primary tool for adaption. What more can reasonably be asked of a real world? (Kuhn 1990, 102).[39]

To conclude, let me indicate briefly where I see Carnap, Quine, and Kuhn diverging in their views. As I see it, Carnap's reconstructive approach of imposing linguistic frameworks on the sciences divides him from Quine and Kuhn and their more naturalistic approaches. By way of pragmatically adopted linguistic frameworks, Carnap introduces a notion of reality that he claims "is an empirical, scientific, non-metaphysical concept" (Carnap 1950/1956, 207). As we have seen, at just this point Quine questions whether this notion of reality is as scientific as Carnap thinks. Why should pragmatically adopting a linguistic framework diffuse the very ontological claims that Quine sees our current best scientific theories making? It seems here that Carnap goes beyond scientific method itself and appeals to "ulterior controls" (Quine 1960, 23). Kuhn, too, takes a naturalistic stance by way of providing an historical account of how the sciences actually develop. And much like Quine, he accepts that scientific theories make genuine ontological claims.[40]

---

[39] Kuhn is reflecting here on his later view, but I think it applies equally well to his earlier view.
[40] Throughout *The Structure of Scientific Revolutions*, Kuhn talks in ontological terms. That he accepts a notion of reality internal to science emerges in his criticisms of those philosophers who demand that a scientific theory's ontology be compared with "what is 'really there'" (Kuhn 1962/1970, 206). See fn. 1 above for accounts of Kuhn that push him much closer to Carnap. Even Friedman, in his (2012), sees Kuhn's ontological focus as distinguishing him from Carnap. Tsou (2015) argues from these recent attempts to push Kuhn and Carnap closer together, emphasizing Kuhn's naturalistic historical approach. Bird (2012), Roth (2020), and Shapin (2015) also emphasize Kuhn's naturalist strand.

I see no reason why we would expect the history of science to show otherwise. Still, as I have argued, each in his respective way has undermined philosophical analyses of reality in traditional realist and anti-realist terms. They are, perhaps, in this sense all Carnapians at heart, rejecting the very substance of the debate between realism and anti-realism as traditionally conceived. Instead, they all aimed to show the continuity between investigations of reality and scientific inquiry. At various times, each has been characterized as a realist or an anti-realist.[41] All are in a sense both and so – echoing a point Burton Dreben made about Quine – neither.[42] This shows how far Carnap, Quine, and Kuhn have shifted the traditional analysis of ontology in terms of realism and anti-realism. The very categories of realism and anti-realism do not apply to their radical reconceiving of philosophy as treating science as the final arbiter in giving an account of what there is.[43]

---

[41] Logical empiricism is commonly taken as an anti-realist philosophy of science. Quine, however, describes Carnap's logicism as realist (1948b, 14). Carnap agrees, so long as realism is understood in the right way (Carnap 1950/1956, 215, fn. 5). For Quine described as one or the other, see fn. 27 above. See also Kemp (2016), especially 177–178. Kemp ultimately argues that Quine is a realist. I think his account squares with the sort of realism Quine espouses. And finally, for Kuhn described as one of the other, see fn. 31 and the beginning of Section 3.3 above.

[42] See fn. 27 above.

[43] I would like to thank Andrew Lugg, Lydia Patton, Erich Reck, and Paul Roth for their comments on an earlier draft of this essay, especially at such short notice. I would like to thank Andrew Lugg additionally for his very detailed editorial comments.

PART II
Carnap, Quine, and American Pragmatism

CHAPTER 4

Pragmatism in Carnap and Quine
Affinity or Disparity?

Yemima Ben-Menahem

4.1 Introduction

When thinking of the role of pragmatism in Quine's philosophy, the famous passage at the end of "Two Dogmas of Empiricism" (1951f) comes to mind:

> Carnap, Lewis and others take a pragmatic stand on the question of choosing between language forms, scientific frameworks; but their pragmatism leaves off at the imagined boundary between the analytic and the synthetic. In repudiating such a boundary I espouse a more thorough pragmatism. Each man is given a scientific heritage plus a continuing barrage of sensory stimulation; and the considerations which guide him in warping his scientific heritage to fit his continuing sensory promptings are, where rational, pragmatic (Quine 1951f, 46).

By "pragmatic considerations" Quine means considerations that are not imposed on us by either logic or experience, considerations that may involve norms and with regard to which we have discretion. That there is room for discretion and norm-guided choice is an immediate result of the underdetermination of theory by observation, one of the core tenets of "Two Dogmas of Empiricism" and Quine's philosophy in general.[1] By analogy with underdetermination in algebra (when there are fewer equations than variables, a situation in which one can choose the values of some variables by fiat and solve the equations for the rest), the underdetermination of theory implies a considerable amount of freedom with regard to theory choice. Underdetermination is visualized in Quine's web of belief, whose inner parts can be variously connected with experience, creating the different options we can choose from. It is a seamless web,

---

[1] See, however, ch. 6 of *Conventionalism* (Ben-Menahem 2006) for a discussion of Quine's critical examination of the thesis of underdetermination.

with no sharp boundary between analytic and synthetic truths, that is, in principle, every component of the web could be revised to achieve better correspondence with experience. The web is not subject to rigid rules of change and modification; when problems come up, there are only "soft" guidelines for choice among empirically equivalent alternatives. Primary among such choice-guiding considerations is Quine's "maxim of minimum mutilation" (1970d, 85), which instructs us to keep most of the current web of belief intact, making changes only where necessary. Quine sees such pragmatic considerations as rational, implying that there could be other, less rational, considerations (e.g., authority) which occasionally affect our choice, but which he would not sanction.

Quine represents his argument as an extension of pragmatism, and yet his referring to a logical positivist such as Carnap in this context speaks against a direct association between the pragmatic argument espoused here and American pragmatism. What Quine means by "pragmatism" in this passage is epistemic discretion, which is common to his own and Carnap's positions. The question of whether, on the basis of this schematic embracement of pragmatism, we can see Quine's philosophy as rooted in the philosophical school bearing that name is tricky, for it is possible that what Quine has in mind is basically an "empiricism without dogmas," as suggested by the title of the final section of "Two Dogmas of Empiricism." Leaving the question regarding Quine's debt to American pragmatism open at this stage, I would like to distinguish the more common use of the term "pragmatic", which can indeed be linked quite naturally to epistemic discretion, from the rubric of pragmatism when associated directly with American pragmatism. Although epistemic discretion was not ignored by the American pragmatists, it is not as central to them as other issues close to their heart and certainly marginal in comparison with the attention devoted to it by Carnap and Quine.

In terms of his biography, Carnap is of course more remote from American pragmatism than Quine. In his case, it is even more important to distinguish the role of discretion and pragmatic considerations in the common sense of "pragmatic" from aspects of his philosophy that could be attributed to his encounter with American pragmatism. Quine was right to ascribe to Carnap the affirmation of discretion in science and philosophy, but I will argue that for Carnap, discretion did not originate in American pragmatism, but rather in *conventionalism*, a position Quine sought to rebut. After becoming acquainted with philosophers subscribing to pragmatism, Carnap, who turned out to be somewhat more generous than Quine in his evaluation of this school (see below), mainly stressed its affinity to his "native" logical positivism. Here, it is quite obvious that Carnap tends to read pragmatism simply as empiricism, a reading that could perhaps be justified by some of the American pragmatists' self-portrayals, but which misses distinctive characteristics of pragmatism that the broad umbrella of empiricism obscures. The question of whether Quine shared this reading is addressed below.

The essay begins with a brief survey of recurring themes in American pragmatism (Section 4.2). In the light of the above distinction between the two senses of pragmatism, Sections 4.3 and 4.4 discuss the place of discretion in Carnap and Quine, while Section 4.5 examines their relation to American pragmatism.

4.2 Pragmatism

Like other philosophical schools, pragmatism cannot be given a precise definition, but I will mention a number of characteristics that recur in pragmatists' writings. It is convenient to construe most of these in negative terms, that is, as pragmatists' disavowal and critique of positions they saw as prevalent in traditional philosophical schools.

No Foundationalism: A major difference between seventeenth-century epistemology and that of the American pragmatists pertains to foundationalism. In the seventeenth century, erasing all previous beliefs and purported knowledge claims so as to make room for a completely fresh start was considered to be the right epistemic method. The underlying metaphor was that of a building – a clean building site, a firm foundation, and a systematic construction ensured the building's strength. Both discovery and justification were thought to be taken care of by this procedure. Peirce, however, was strongly opposed to the seventeenth-century recipe. Whatever the firm foundation was supposed to be, whether it consisted of self-evident truths or bare sense data, the foundationalist method could not work. Peirce's objection to Descartes, on the one hand, and the empiricists, on the other, was that one could neither destroy the entire body of previous belief nor construct a new one from scratch. Instead, investigation always begins with a localized problem, an island of doubt in a sea of stable and undisputed belief.[2] Similarly, James stressed our dependence on a vast amount of traditional knowledge. A total revision of the system cannot even be conceived: "An *outrée* explanation, violating all our preconceptions, would never pass for a true account

---

[2] This critique appears in several of Peirce's writings, but see in particular his (1868).

of a novelty …. The most violent revolutions in an individual's beliefs leave most of his old order standing" (James 1907/1955, 50–51).

No Skepticism: Skepticism, even if only a methodological point of departure, as in Descartes, and certainly when adopted as a sustained epistemic position, is an anathema to pragmatists. The space drained of belief, they claim, the space in which skeptics presume to be living in a blissful state of suspended judgment, is unfit for human beings. Skeptics may be right to hold that perfect justification is impossible, but, pragmatists retort, so is global doubt. Even local doubt, though, cannot be triggered at will, out of blind obedience to a methodological maxim. It must be a "living doubt."

> Some philosophers … recommended us to begin our studies with questioning everything! But the mere putting a proposition in the interrogative form does not stimulate the mind to any struggle after belief. There must be a real and living doubt, and without all this, discussion is idle. (Peirce 1877/1966, 100)

Moreover, it is often assumed that belief must be justified, whereas doubt needs no reasons. By contrast, pragmatists maintain that doubt too requires justification. James argued further that in some cases belief beyond evidence is justified – his notorious "will to believe" or "right to believe" (James 1897/1956). The rejection of both skepticism and foundationalism go hand in hand in recognizing that we can only address localized problems set against a background of what is taken for granted. Even though this background is neither intrinsically beyond doubt nor everlasting, it is "good enough" to guide us toward further knowledge. The provisional status of background beliefs allows pragmatists to reject skepticism while, at the same time, endorsing fallibilism.

Fallibilism: Pragmatists repeatedly stressed that there could be no incorrigible beliefs. While we cannot revise the entire system of knowledge at once, we must often revise some of its components. Due to interconnections between components of the system, revision of one part may eventually affect others, but we can never revise the whole system at once. Revisionism applies to any fraction of the system, from observation reports to logical and mathematical truths.

No Essentialism: The pragmatist theory of meaning, in both its Peircean and Jamesian versions, stands in marked contrast to essentialism. Indeed, it was the aim of these thinkers to develop an alternative to traditional conceptions on which meanings consist in fixed essences that our definitions strive to capture. The alternative was the dynamic and empiricist approach to meaning that has become the emblem of pragmatism.

Blaming essentialism for a series of philosophical blunders, James finally condemns it as no less than a form of magic.

> Metaphysics has usually followed a very primitive kind of quest. You know how men have always hankered after unlawful magic, and you know what a great part, in magic, *words* have always played. If you have his name, or the formula of incantation that binds him, you can control the spirit, genie, afrite, or whatever the power may be. ... The universe has always appeared to the natural man as a kind of enigma, of which the key must be sought in the shape of some illumination or power-bringing word or name. That word names the universe's principle, and to possess it is ... to possess the universe itself. (1907/1955, 46, italics in original)

<u>The Social Dimension of Knowledge</u>: In the seventeenth century, epistemology was generally conceived in terms of the mental activity of individuals. Pragmatists, on the other hand, stress the social character of language and knowledge. Recall Peirce's theory of signs. In addition to the sign and the signified, there is always an *interpreter*, who can be an immediate addressee as well as a future one. Meaning is generated in the prolonged interaction between speakers and interpreters. Likewise, the creation of knowledge is a long-term social process carried out by a community of investigators. The need for a linguistic and epistemic tradition, and the fact that verification is a long-term process to which many individuals contribute, is also acknowledged by James.

<u>Belief and Action</u>: Pragmatists maintain that belief manifests itself in action. Inspired by Alexander Bain's 1859 *The Emotions and the Will*, this conception is appealing to pragmatists in its focus on concrete, observable consequences, rather than obscure mental states. Although central to American pragmatism, the connection between belief and action is marginal in Carnap and Quine, and thus in what follows.

<u>No Correspondence Theory of Truth</u>: James ridicules the notion of an "absolute correspondence of our thoughts with an equally absolute reality" (1907/1955, 54). The world is given to us in language – a human creation. We compare our various descriptions of the world with one another, but we cannot compare them with reality, pre-linguistic fact, or a thing in itself. Given the intersubjective nature of language, dependence on language does not make our knowledge claims subjective in the pejorative sense of being up to the individual's whim. Direct correspondence is denied, but one should not conclude that anything goes. Peirce and Dewey have also argued against the correspondence theory, and against the naïve realism it gives rise to. Despite their agreement on the flaws of the correspondence theory, however, pragmatists vary significantly on the positive accounts

they offer in its stead. The "pragmatic theory of truth" is therefore a misleading term. Peirce defined truth as "the opinion which is fated to be ultimately agreed to by all who investigate" (1878/1966, 133). And he took the term "fate" seriously, though not in the sense of reifying fate or seeing it as a voluntary decree by a conscious being: "This activity of thought by which we are carried, not where we wish, but to a foreordained goal, is like the operation of destiny" (Ibid.). In contrast to Peirce, James had a pluralistic conception of truth: he distinguished verifiable from unverifiable beliefs, well-rooted from novel ones, those on which we can postpone our verdict from those that must be urgently decided on, and those that are independent of our belief in them from those that are up to us to *make* true by believing in them (as when a person who believes she will succeed stands a better chance of doing so than a person believing she is bound to fail).

Rethinking Traditional Dichotomies: Dualities such as the analytic/synthetic, the fact/value, and the meaningful/meaningless dyads have been an integral part of the empiricist tradition. Although the American pragmatists were inspired by that tradition, they expressed reservations about the cogency of these strict dualities. Pragmatists keen to address moral and social issues, such as James and Dewey, found the fact/value dichotomy particularly objectionable since it implied that there could be no rational deliberation of moral issues. In general, an overly restrictive account of meaning was considered by the pragmatists repugnant on both epistemic and moral grounds. Disapproval of such dichotomies turned out to be a matter of dispute between the pragmatists and the logical positivists and, in particular, between Carnap and Quine.

Before trying to detect traces of the influence of this pragmatic heritage on Carnap and Quine, I turn to the more limited sense of "pragmatic" employed when pointing to the need of pragmatic considerations in cases of underdetermination and epistemic discretion.

4.3 Discretion in Carnap

The principle of tolerance, which promotes "complete liberty with regard to the forms of language" (1934/1937, xv) and asserts that "in logic, there are no morals" (1934/1937, 52), constitutes Carnap's best known declaration of freedom and discretion. The principle culminated a long journey during which Carnap struggled with questions regarding indeterminacy and discretion in a number of philosophical areas: the geometrical structure of space, the meaning of concepts, and the controversy about the foundations of mathematics.[3] In all of these areas Carnap was concerned

---

[3] In this controversy, Carnap was mainly interested in two of the alternatives, logicism and formalism.

with the pros and cons of conventionalism. Carnap is explicit about the connection between his principle of tolerance and conventionalism, introducing the principle thus: "It is not our business to set up prohibitions, but to arrive at conventions" (1934/1937, 51). Reflecting on the principle in later years, he remarked that a more appropriate name for it would have been "the principle of conventionality" (1942, 247) or "the principle of the conventionality of language forms" (1963a, 55).

Carnap's engagement with conventionalism is already evident in his doctoral dissertation on space, published as *Der Raum* (Carnap 1922a), and continued to manifest itself in the following decade in his various works on the philosophy of science and the foundations of logic and mathematics. Though the dissertation makes just a few explicit references to Poincaré, the influence of his writings is conspicuous. Carnap accepted Poincaré's threefold division of space into formal, physical, and intuitive (the latter parallels Poincaré's representational space) and argued for the conventionality of the metric of physical space on the basis of its being underdetermined by experience.

Der Raum exhibits familiarity with Hilbert's *Grundlagen der Geometrie* (Hilbert 1899) and its notion of axioms as implicit definitions. Carnap notes that the axioms and theorems of formal space (in contrast to those of intuitive and physical space) do not presuppose any particular meaning of the primitive symbols (1922a, 3). The formal approach of *The Logical Syntax of Language* (Carnap 1934/1937; henceforth LS) similarly presupposes Hilbert's method. Indeed, Herbert Feigl (1975, xvi) suggested to Carnap that the basic idea of LS amounts to a "Hilbertization" of *Principia Mathematica*, a suggestion Carnap apparently found agreeable. However, in 1963, Carnap distinguished between Hilbert's syntactic method, which he had endorsed and Hilbert's philosophy – formalism – which he had rejected (Carnap 1963b, 928). Carnap's efforts to come to terms with the idea of implicit definition and its bearing on the problem of meaning are evident in an interesting and somewhat neglected paper, "*Eigentliche und uneigentliche Begriffe*" (Determinate and Indeterminate Concepts) (Carnap 1927). As the title indicates, the paper addresses the question of whether concepts have definite meanings, a question that is particularly pressing with regard to formal systems, which by their nature are amenable to multiple interpretations. Moreover, according to the Löwenheim–Skolem theorem, formalisms rich enough to contain arithmetic are noncategorical, that is, they are bound to have nonisomorphic models.

In the paper, Carnap treats the issue of categoricity under the rubric of the "monomorphism" or "polymorphism" of formal systems. He begins by presenting a Fregean characterization of concepts (*Begriffe*) as functions that objects (or classes of objects) determinately satisfy or fail to satisfy. He further distinguishes real concepts, which refer directly to physical reality, from formal concepts, the concepts of logic and mathematics, which do not designate real entities but are nonetheless essential for speaking about reality. Regarding the former, he expresses the conviction, which he promises to substantiate in his forthcoming *Aufbau* (Carnap 1928a), that the entire corpus of knowledge – including even psychology, sociology, and the history of religion – can be systematically constituted (by way of a *Konstitutionstheorie*) from a very small number of physical concepts. Moving on to formal concepts, numbers in particular, Carnap compares Russell's explicit definition of the natural numbers with their implicit definition via Peano's axioms. The disadvantage (*Nachteil*) of the latter definition is that it is multiply interpretable. Peano's axioms define a progression – a recursive structure with an infinite number of (formal and informal) applications or realizations (*Anwendungen*). Carnap illustrates that when an axiom system is polymorphic, that is, when its models are nonisomorphic, there are questions that receive different answers in different models, and concepts that are indeterminate – they are applicable in some models, but not in others.

    Carnap is struggling with the problem that was at the heart of the Frege–Hilbert controversy. If concepts must be determinate, as Frege insisted, and if implicitly defined concepts are typically indeterminate, we must reconsider their status as concepts. Indeed, Carnap agrees with Frege that implicitly defined "concepts" are actually variables, and that the "theorems" in which they appear are only theorem schemata. But this deficiency does not impel Carnap to dismiss implicit definition altogether. For one thing, even when an axiomatic system functions as an implicit definition of its terms, it always provides an *explicit* definition of a type of structure. Peano's axioms define the natural numbers implicitly, and the notion of a progression explicitly. Carnap perceives this feature of implicit definitions as an opening that will allow the views of Frege and Hilbert to be reconciled, a solution that had eluded both of them. For another, the crucial consideration pertaining to definitions is their *fruitfulness* (*Fruchtbarkeit*), rather than their truth. At this juncture, Carnap embarks on a conventionalist route and parts company with Frege. Theoretically, there is no way to narrow down the number of admissible interpretations of a polymorphic system of axioms, but admissible interpretations can differ significantly in terms of their fruitfulness, as becomes evident when a formalism is brought into contact with reality by means of a "realization" – an empirical interpretation of its terms. Without such a realization, implicitly defined concepts "hang in thin air" (*schweben in der Luft*), but when contact with reality is established – and here Carnap uncharacteristically resorts to figurative language – "the blood of empirical reality flows ... into the veins of the hitherto empty schemata, thereby transforming them into a full-blown theory" (1927, 372–73). The multi-interpretability of formal systems needs no longer worry us; on the contrary, in allowing for new possibilities for realization, multi-interpretability enhances fruitfulness. It is only by being anchored in empirical reality that formalisms turn into theories, which are potentially true or false. Here we can detect the formal account of logic and mathematics that is at the center of LS, with its emphasis on empirical applications of these areas. But in LS, the distinctions between form and content, structure and theory, and convention and truth are further elaborated by means of tools that Carnap had not yet come up with in "Eigentliche und uneigentliche Begriffe," principally, the distinction between the material and the formal modes of speech.[4]

Another, more familiar, path from Carnap's early work to his mature philosophy takes us to the debate among logicists, formalists, and intuitionists on the foundations of mathematics. The significance of this debate for the development of the principle of tolerance has been frequently noted, including Carnap's own remark on that issue in his autobiographical notes (1963a). Here, too, Carnap came to believe that there was no one correct position, but only conventional choices to be made in the light of pragmatic interests. Although Carnap tells us that the spirit of tolerance had always guided his thinking on ideological issues, there is a marked difference between the tolerance reflected in his earlier attempts at reconciling conflicting views on the foundations of mathematics, and that mandated by the fully articulated principle of tolerance. In the papers written before LS, Carnap, though hopeful about the prospect of a truce, definitely leans toward logicism. Thus, his quasi-logicist "*Die Mathematik als Zweig der Logik*" (1930) optimistically concludes that while the problems of the foundations of mathematics have not yet been fully resolved, a more peaceable coexistence between formalism and logicism is at hand. A couple of years later, this possibility would be realized – it becomes firmly anchored in LS and its principle of tolerance.

---

[4] According to Awodey and Carus (2007), the principle of tolerance was not yet part of the early drafts of *Logical Syntax*. In my *Conventionalism* (Ben-Menahem 2006), I argued that one of Carnap's major goals was to retain what he could from the *Tractatus*, while overcoming what he saw as its difficulties, for example, the say/show distinction. Inspiration from both Hilbert and Gödel was crucial here. In their profound paper, Awodey and Carus provide a similar account in greater detail.

I have surveyed the prehistory of Carnap's principle of tolerance at some length, both because the early writings (not all of which have been translated into English) are not as well-known as LS (and subsequent writings) and because they point very clearly to the origin of Carnap's concern over epistemic discretion in geometry and mathematical logic, areas in which conventionalism was accumulating interest and credibility. In Carnap's later work – and here I can afford to be brief – discretion continues to occupy central stage. Not only in LS, but also in "Empiricism, Semantics, and Ontology" (Carnap 1950/1956), where he introduces his celebrated distinction between internal and external questions, Carnap proclaims the freedom to choose linguistic frameworks on the basis of pragmatic considerations such as utility or convenience. He guards against the conflation of theoretical questions, which can be given true or false answers, and practical questions that require "a practical decision concerning the structure of our language" (1950/1956, 207).

Two points need to be stressed. First, Carnap thinks of the above distinctions, the material versus the formal modes of speech, external versus internal questions, the theoretical versus the practical, and truth versus convention as dichotomies, allowing no vagueness or borderline cases. Secondly, the alternatives in question are visible and the choice among them is a genuine choice, that is, it is conscious and voluntary. One could of course be in error, addressing, for example, a practical question as if it were a theoretical one. But in principle, once the distinctions are understood, there is no reason why one would not be fully aware of the alternatives one confronts and the considerations one should take into account when making one's choice. Both of these characteristics of Carnap's position were unacceptable to Quine.

4.4 Discretion in Quine

In *Word and Object* (1960), Quine moves from the underdetermination of theory espoused in "Two Dogmas of Empiricism" to the indeterminacy of translation. The argument for indeterminacy is first developed for "radical translation," namely, for a field linguist who tries to learn a completely foreign language, but it soon becomes evident that the argument applies just as well to one's own language. What is therefore at issue is not just the indeterminacy of translation, but the indeterminacy of meaning in general. What exactly is the relation between the underdetermination of theory and the indeterminacy of translation? Quine addresses the question explicitly in "On the Reasons for Indeterminacy of Translation"

(1970c). He identifies two reasons for indeterminacy: the first is closely related to the underdetermination of theory and affects the truth of sentences; the second – "the inscrutability of reference" – is independent of the former and asserts that sentences do not determine their reference uniquely. The first kind of indeterminacy arises because, when language learners try to figure out what their fellow speakers mean, they go by these speakers' overt linguistic behavior, which they can passively watch or actively trigger by queries. On the basis of this data, learners create a theory about the meaning of utterances they encounter, or, as Quine puts it (avoiding the term "meaning"), a translation of these utterances into their native language. Quine argues that this theory/translation is underdetermined by the data gathered from linguistic behavior of speakers in the same way and to the same degree that scientific theories are underdetermined by observation. "What degree of indeterminacy of translation you must then recognize … will depend on the amount of empirical slack that you are willing to acknowledge in physics" (1970c, 181).

The second argument for indeterminacy, explored in detail in "Ontological Relativity" (1969c), arises when instead of looking at the evidence and asking whether there is more than one theory compatible with it, one turns to sentences (or theories) and raises the question of whether they uniquely determine what they are *about*. Quine's reply is negative: agreement on the truth-values of sentences does not entail agreement on ontology. He offers two kinds of arguments to support this second kind of indeterminacy: The first is an informal argument to the effect that ontology presupposes individuation schemes that cannot be extracted from the raw data of speakers' dispositions, but must be imposed upon the data by the interpreter. The second draws on the Löwenheim–Skolem theorem to reach the same result. As mentioned, the theorem proves that theories rich enough to include arithmetic have an infinite number of different models, not all of them isomorphic to one another. It follows that to accept such a theory as true does not guarantee that we know what the theory is about. Theory (of the appropriate kind) fixes neither reference nor ontology. Quine sums up:

> There are two ways of pressing the doctrine of indeterminacy of translation to maximize its scope. I can press from above and press from below, playing both ends against the middle. At the upper end there is the argument … which is meant to persuade anyone to recognize the indeterminacy of translation of such portions of natural science as he is willing to regard as underdetermined by all possible observations. … By pressing from below I mean pressing whatever arguments for indeterminacy of translation can be based on the inscrutability of reference. (1970c, 183)

Returning to discretion, do Quine's arguments for the indeterminacy of translation imply the same kind of discretion he ascribed to the scientist in "Two Dogmas of Empiricism"? Does Quine suggest that speakers confront genuine choices of the kind Carnap offered to those worried about the existence of numbers or physical objects? Recall that Quine's arguments for indeterminacy were not meant to prove failure of communication, but to critique a theory of meaning that Quine had considered naive – "uncritical semantics" as he calls it in "Ontological Relativity" (1969c, 27). "Uncritical semantics is the myth of a museum in which the exhibits are meanings and the worlds are labels" (Ibid.). The assumption underlying this myth is that "a man's semantics [is] somehow determinate in his mind beyond what might be implicit in his dispositions to overt behavior" (Ibid.). The thrust of the arguments for indeterminacy is therefore iconoclastic rather than skeptical. Meaning as a reified entity is an idol, but smashing the idol does not deprive us of speech and communication. "The word 'meaning' is indeed bandied as freely in lexicography as in the street, and so be it. But let us be wary when it threatens to figure as a supporting member of a theory. In lexicography it does not" (1995a, 83).

If the iconoclastic reading of Quine's arguments is correct, then underdetermination and indeterminacy are not invariably accompanied by conscious discretion and genuine choice. In "Two Dogmas of Empiricism," it appears, they are, for there Quine envisages a scientist aware of empirically equivalent alternatives, one of which is pragmatically preferable to her and which she decides to use. But in *Word and Object*, it is less clear that an informed decision can always be made. Even in Quine's paradigmatic "Gavagai," the linguist is not expected to be fully aware of the various options and to make a conscious decision between "rabbit," "undetached rabbit parts," and so on. Quine is quite explicit about that, noting that although translation always involves some scheme, some "analytical hypothesis," which guides the translation, this hypothesis may well be unconscious (1960, 71). On this point there is a significant difference between Carnap and Quine. While both of them draw attention to epistemic gaps in various areas of science and philosophy, and in linguistic frameworks, and while they acknowledge the role of pragmatic considerations in bridging some of these gaps, they nonetheless differ in their understanding of the decision process. Carnap's decision maker surveys the alternatives from above, as it were, having all of them in clear view and making a rational and voluntary choice among them. By contrast, Quine sees us as always working locally, from within our own system and with no access to Carnap's panoramic viewpoint. This feature of Quine's philosophy is closely linked to his immanent notion of truth, to which I return later, and to his view of philosophy as continuous with science.[5] "The philosopher's task differs from others', then, in detail; but in no such drastic way as those suppose who imagine for the philosopher a vantage point outside the conceptual scheme that he takes in charge. There is no such cosmic exile" (1960, 275).

A much more explicit disparity between Carnap and Quine developed over the dichotomies mentioned at the end of the previous section. Quine's critique of Carnap's dualities is already noticeable in his three Harvard lectures on LS (Quine 1934), and recurs in many other places.[6] One of these, "On Carnap's Views on Ontology," combines deference and critique: "Though no one has influenced my philosophical thought more than Carnap, an issue has persisted between us for years over questions of ontology and analyticity" (Quine 1951a, 203). Targeting "Empiricism, Semantics, and Ontology" (Carnap 1950/1956), Quine argues that Carnap's distinction between external and internal questions boils down to the distinction between analytic and synthetic truths, a distinction he had subjected to criticism in "Two Dogmas of Empiricism." Having compared Carnap and Quine's views on the specific issue of epistemic discretion, we can now address the more general question regarding the possible connections between their philosophies and American pragmatism.

4.5 Carnap and Quine vis-à-vis American Pragmatism

The development of Carnap's thinking follows a trajectory of liberalization with regard to both the observational basis of science and the logic of the language of science. One of the peaks of this journey is the principle of tolerance cited above, but there were earlier steps in the direction of tolerance and pluralism. Such was the transition from the phenomenalist understanding of the observation basis espoused in the *Aufbau* to its physicalist understanding in the early 1930s (1932/1987), and the transition from a search for a firm foundation of mathematics to the pluralistic position Carnap was striving to

---

[5] Quine was less decisive on the issues discussed than this passage suggests. As he admits, he wavered between what he called the "sectarian position" (which I have ascribed to him here) and the "ecumenical position." I have described his vacillation at length elsewhere (Ben-Menahem 2006, ch. 6).

[6] On the face of it, the general spirit of the lectures is much more appreciative of Carnap's book than "Truth by Convention," written only two years later (Quine 1936). In an attempt to harmonize the two, Richard Creath (1990a) reads "Truth by Convention" as less critical of Carnap than is usually thought, and more in line with the lectures. By contrast, I have argued that although the rhetoric is different, the main ingredients of the critique launched in "Truth by Convention" are already part of the lectures (Ben-Menahem 2006, ch. 6).

develop in the (1927) and (1930) papers I have described. Moreover, the evolution toward a more liberal empiricism and a more pluralist conception of meaning did not stop with LS. "Testability and Meaning" (Carnap 1936a and 1937) and "Empiricism, Semantics, and Ontology" (1950/1956) are further milestones in the evolution of Carnap's philosophy. Carnap's early writings show hardly any interest in, or familiarity with, the writings of American pragmatists. In 1933–34, this indifference gave way to rather elaborate intellectual relations with a number of American philosophers – Nagel, Morris, Lewis, Quine – established either through their visits to Europe or through correspondence. Carnap's move to the United States in 1935 further deepened these relations. The question of whether the newly acquired acquaintances had an impact on the trajectory of Carnap's thinking naturally arises. Since Carnap's later views are indeed closer to those of the American pragmatists than his earlier ones, one tends to answer this question in the affirmative. Limbeck-Lilienau (2012) has argued convincingly, however, that some of the moves that could be interpreted as Carnap's response to the pragmatists' critique of logical positivist positions had in fact been made earlier than his encounter with pragmatism and cannot be ascribed to the influence of this encounter. Specifically, by 1933 Carnap had already distanced himself from the phenomenalist/individualist/reductionist position of the *Aufbau*, which was the target of pragmatists' critique, now embracing physicalism, holism, the social character of science, and the limits of verification. Although these shifts in Carnap's position were independent of his new American contacts, and although the interaction between Carnap and the pragmatists began with some misunderstanding (on the part of the pragmatists) as to Carnap's actual views at the time, the groundwork for a philosophical alliance had been laid.

Beyond the philosophical details of the growing rapprochement between the two schools, one must remember that the logical positivists (and their allies in Warsaw and Berlin) saw themselves as a minority whose views were at odds with the prevailing philosophical atmosphere. Under these circumstances, the alliance with American pragmatism must have had an emotional effect that went deeper than mere intellectual accord. With the rise of Nazism in Europe, the friendship of American colleagues became even more significant. Given the dissimilarity in their social/political situation, it is only natural that an immigrant such as Carnap was more eager to harmonize the opinions of the two schools than an American philosopher such as Lewis. With the growing influence of Carnap (as well as other immigrants, such as Reichenbach) in the United States, one could perhaps even identify some resentment, on the part of "native" American philosophers such as Lewis, toward this successful logical-empirical "invasion."

We have seen that Carnap was deeply engaged with conventionalist ideas both in his philosophy of science and in his philosophy of logic and mathematics. Conventionalism is based on the recognition that not every sentence that appears to express an assertion (and therefore appears to have a determinate truth value) does in fact do so. On the conventionalist understanding, the axioms of geometry provide a clear example of this kind of camouflage. Their formulation as assertions is misleading; in fact, they are definitions in disguise, that is, they are neither true nor false. Conventionalists saw the distinction between truth and convention as vital. In their view, it was a major philosophical desideratum to separate fact from convention, objective truth from the language chosen to express it. Carnap, early and late, respected this distinction. His "American" writings continue to pursue ways of distinguishing what is up to us – linguistic frameworks – from what is imposed on us by empirical or mathematical reality. In other words, he remained faithful to the conventionalist tradition. Thus, even his later writings show more kinship to the tradition that inspired him in his formative years than to the tradition of American pragmatism that he met as a mature philosopher.

One would assume that Quine's links with American pragmatism are stronger than Carnap's, but Quine's writings on pragmatism do not support this conjecture, at least not explicitly. We will first examine these writings and then argue that the impact of American pragmatism on Quine's thinking is in fact deeper than his papers on pragmatism suggest. In 1975, Quine was invited to deliver a paper at a conference on "The Sources and Prospects of Pragmatism." His paper appeared in 1981 in two versions: the full version, entitled "The Pragmatists' Place in Empiricism," in a volume containing the conference papers (Quine 1975), and an abridged version, under the name "Five Milestones of Empiricism," in *Theories and Things* (Quine 1981a). The latter contains only part of the full version, excluding any discussion of pragmatism. The fact that, in his collection, Quine chose to include only the sections on the five milestones of empiricism indicates more than anything that he said openly that he ascribed little significance to the impact of pragmatism on his own thought, and was perhaps also doubtful about the importance of pragmatism in general. The reader gets the same impression from what Quine does say about pragmatism in the unabridged version of the paper. His main complaint is that "it is not clear ... what it takes to be a pragmatist .... [T]he term 'pragmatism' is one we could do without" (1975, 23). What pragmatists share, according to Quine, is empiricism, even if not the specific brand of empiricism he commends. Hence, the "five points where empiricism has taken a turn for the better" (1975, 23), constituting the milestones on the road to Quine's own empiricism, are as follows:

1. The shift from ideas to words.
2. The shift from terms to sentences.
3. Holism – the shift from sentences to systems of sentences.
4. No analytic-synthetic dualism.
5. Naturalism – no prior philosophy.

Quine ascribes the first of these transitions to John Horne Tooke's critique of Locke and the second to Jeremy Bentham. The founding fathers of analytic philosophy, Frege, Russell, and Wittgenstein, as well as the logical positivists, are also mentioned by Quine as promoting these two insights. The remaining three transitions are characteristically Quinean even if he deemphasizes his role as their proponent; for example, he cites August Comte as the originator of naturalism. Notably, none of the five points is attributed by Quine to the pragmatists.

In the ensuing discussion of the position of pragmatists on his five points, Quine mentions various disagreements with them. He criticizes Peirce for vacillating between words and ideas, and between beliefs and sentences (although ultimately settling for sentences), and for not being sufficiently outspoken about holism. He criticizes James for being kind to wishful thinkers and both James and Dewey for declining (immanent) realism. Finally, he disagrees with Lewis about the analytic-synthetic distinction. Quine mentions in passing a couple of points of agreement with the pragmatists: fallibilism, the repudiation of Cartesian doubt, and the recognition of Darwinism as a key to understanding the human mind and its conceptual categories. There are two further points for which Quine credits the pragmatists somewhat more willingly: the man-made nature of truth and the social character of meaning. The latter he refers to as "behavioristic semantics", stressing that it was Dewey, rather than Wittgenstein, who first insisted "that there is no more to meaning than is to be found in the social use of linguistic forms" (1975, 36–37). The concluding lines of Quine's generally critical paper on pragmatism are more favorable than its opening ones, but it remains a fact that Quine saw none of the five advances in empiricism as initiated by pragmatism, and that he omitted the entire discussion of pragmatism from the version included in his collection.

Quine's somewhat dismissive attitude notwithstanding, he was apparently more deeply rooted in the American pragmatist tradition than his paper, in both the full and the abridged versions, recognizes. Don Howard (2018) explored some of these roots, stressing in particular the affinity between Quine and Dewey, an affinity Quine acknowledged in his John Dewey Lectures published in (Quine 1969c). More surprising, perhaps, are the similarities between Quine and James, a philosopher Quine hardly mentioned. Here are some examples.

To begin with, James is committed to empiricism, a commitment evident not only in *Pragmatism* (1907/1955), but also in *The Principles of Psychology* (James 1890), his pioneering attempt to turn psychology into an empirical science. His perception of pragmatism as continuous with empiricism is reflected in his portrayal of pragmatism as "a new name for some old ways of thinking" (the subtitle of *Pragmatism*). The principal epistemic desideratum, in his view, is conformity with experience. "But this all points to direct verifications somewhere, without which the fabric of truth collapses, like a financial system with no cash-basis whatever. … Beliefs verified concretely by somebody are the posts of the whole superstructure" (1907/1955, 52). And again, "But all roads lead to Rome, and in the end and eventually, all true processes must lead to the face of directly verifying sensible experiences somewhere" (1909/1955, 141). The distance between this firm commitment to empirical support and the common image of James as sanctioning irresponsible make-believe should be obvious.

Secondly, like Quine, James opposed the analytic-synthetic dichotomy. What traditional philosophers saw as eternal and incorrigible truths were for James only "the dead heart of the living tree" (1907/1955, 53). Not completely dead, however, for "how plastic even the oldest truths nevertheless really are has been vividly shown in our day by the transformation of logical and mathematical ideas, a transformation which seems even to be invading physics" (1907/1955, 53). The theorems of logic and mathematics and even some of the laws of nature were once conceived as representing "the eternal thoughts of the Almighty. His mind also thundered and reverberated in syllogisms. He also thought in conic sections, squares and roots and ratios, and geometrized like Euclid" (1907/1955, 48). In fact, however (James argues), all of these laws "are only a man-made language, a conceptual shorthand … in which we write our reports of nature, and languages, as is well known, tolerate much choice of expression and many dialects" (Ibid., 49). James's reason for seeing the system as man-made is the same as Quine's, namely, that language, with its categories and classifications, is a human creation. There is no privileged language that can be singled out as a true description of reality, no language that nature should have used to describe itself, so to speak. "The trail of the human serpent is thus over everything" (1907/1955, 53). The similarity with Quine is also manifest in James's account of the dual traffic between theory and experience. Whereas we typically change theory to accommodate recalcitrant experiences, both Quine and James also countenance the reverse process, whereby we sacrifice an observation sentence (or reinterpret it) in order to save parts of our theory. The feasibility of this option speaks against a simplistic picture of observation as a secure basis to which every theoretical sentence can be reduced. "New truths thus are resultants of new experiences and of old truths combined and mutually modifying one another" (1907/1955, 113).

Finally, and most strikingly, James is explicit about underdetermination, discretion, and the criteria that are involved in the decision on the preferred theory. "Yet sometimes alternative theoretic formulas are equally compatible with all the truths we know, and then we choose between them for subjective reasons ... taste included, but consistency both with previous truth and with novel fact is always the most imperious claimant" (1909/1955, 142). When faced with a problem, a tension discovered within the existing system, or a new experience that seems hard to fit into it, one should try to save what one can from the old system, "for in this matter of belief we are all extreme conservatives" (1907/1955, 50). The reasonable method thus aims at "a minimum of disturbance," or "a minimum of modification" (Ibid.). The idea is not only identical with that of Quine in terms of substance; it also uses the same terminology. "New truth ... marries old opinion to new fact so as ever to show a minimum of jolt, a maximum of continuity. We hold a theory true just in proportion to its success in solving this 'problem of maxima and minima'" (1907/1955, 50–51).

Despite these striking similarities between James and Quine, there are also differences between them, the most significant of which pertains to the notion of truth. Neither James nor Quine were willing to follow Peirce in this teleology of truth. Quine notes (1975, 31) that we have no way of comparing theories in terms of their similarity to one another or in terms of their distance from the truth, a comparison he takes to be presupposed by Peirce's definition. On this point James and Quine are once more in agreement. But Quine was just as opposed to James's ideas about truth as he was to Peirce's. He characterizes his own concept of truth as immanent: "we are always talking within our going system when we attribute truth; we cannot talk otherwise" (1975, 34).[7] It is our best scientific theory that tells us what is true and what is real. To accept this theory and still to refuse to acknowledge its truth (or the reality of the entities it invokes) is senseless, according to Quine. The immanence of truth goes hand in hand with "unregenerate realism, the robust state of mind of the natural scientist who has never felt any qualms beyond the negotiable uncertainties internal to science" (1975, 28). Consequently, for Quine, "physical objects are real, right down to the most hypothetical of particles, though this recognition

---

[7] The connection between Quine's naturalism and his immanent concept of truth is at the center of Verhaegh (2018)

of them is subject, like all science, to correction" (1975, 33). Quine links his immanent conception of truth to his holism:

> The naturalist philosopher begins his reasoning within the inherited world theory as a going concern. He tentatively believes all of it, but believes also that some undistinguished portions are wrong. He tries to improve, clarify, and understand the system from within. He is the busy sailor adrift on Neurath's boat. (Quine 1981a, 72; 1975, 33)

To take stock, both Carnap and Quine understand pragmatism mainly as a version of empiricism and it is under this interpretation that they can sympathize with it. And yet, even where they agree on the general contours of the picture, in particular when both of them make room for epistemic discretion, the origin of their shared position is different. Carnap is inspired by conventionalism, with its firm division between fact and convention, whereas Quine opposes this dichotomy, construing the borders between the two categories as flexible and dynamic. While there seems to be no significant influence of American pragmatism on Carnap, Quine's liaison with that tradition is much more intricate. We have seen that James and Quine share important features of their epistemology: both of them see our belief system as man-made and interconnected, responsible to experience and yet underdetermined by it. The system is seamless in the sense that it has no privileged truths, dynamic in allowing variation of each one of its components, and rationally adjustable when obeying the maxim of minimum mutilation. As far as I know, Quine nowhere mentions the similarity between his ideas and those of James. According to his scientific autobiography in the *Library of Living Philosophers* volume, however, James's *Pragmatism* was one of the only two philosophy books Quine read as a teenager. "I read them compulsively and believed and forgot all" (1986a, 6).

CHAPTER 5

Objectivity Socialized

James Pearson

*Objectivity is a subject's delusion that observing can be done without him.
Invoking objectivity is abrogating responsibility, hence its popularity.*
Heinz von Foerster, Von Foerster and Poerksen (2002, 148).

Despite their fundamental disagreement about the viability of the analytic/synthetic distinction, there is much about which Rudolf Carnap and W. V. Quine agree.[1] Both are empiricists who maintain that our knowledge of the world is wholly based in experience. Both are convinced that formal methods should be used to refine ongoing scientific inquiry. And both reject the idea of supra-scientific philosophical insight into the nature of reality. Recent scholarship defending Carnap from Quine's criticisms has urged that his program is a viable alternative to Quine's, rather than a defeated historical curiosity.[2] But those who take a side in this dispute are typically willing to accept their basic orientation, happy to inherit the tradition of scientific philosophy.

This essay examines an objection that both Quine and Carnap face, namely, that by privileging individual epistemic subjects, they distort the social nature of inquiry. This objection is at the heart of Donald Davidson's claim that Quine fails to grasp the significance of the concept of truth. In Carnap's case, the objection may be detected in Charles Morris's call to ground scientific philosophy in semiotics, the science of signs, rather than syntax, the formal investigation of languages. Drawing out the challenge

---

[1] Indeed, Quine thought that in questioning the analytic/synthetic distinction he was "being more Carnapian than Carnap," since he was attempting to rigorously assess its scientific credentials (1994b, 154).
[2] While some believe that heeding Quine's call to naturalize epistemology purifies empiricism from dogma, others argue that embracing Quine's holism elides distinctions upon which working scientists rely, and that what Michael Friedman has called Carnap's explication of the *a priori* in terms of those truths held fundamental to shaping our current language may better model the significance we award different kinds of revisions to our theories. For discussion, see Friedman (1999) and Ebbs (1997).

from Morris's proposal will require examining a neglected influence on this neglected philosopher: his advisor George Herbert Mead's social theory of mind. I shall argue that Morris and Davidson can both be understood as issuing a demand that scientific philosophers socialize their conception of objectivity.

In the first section, I present Morris's program of scientific empiricism as not just a bridge-building tactic for bringing pragmatists and logical positivists together but as actively demanded by the theory of mind he takes from Mead. Morris wants to help logical positivists recover from what he calls their "individualist hangover" (1937, 24), and fears that Carnap's lack of attention to the communal nature of scientific endeavors threatens to undermine his formal approach. In the second section, I contrast Morris's scientific empiricism with Quine's naturalism. Quine seeks to embed philosophical inquiry within scientific practice, and the new, intersubjective conception of objectivity he crafts at first seems capable of meeting Morris's concerns. Nevertheless, in the third section, I turn to Davidson's criticism that Quine fails to capture the "objective character" of thought (2001b, 10). Like Mead and Morris before him, Davidson thinks that epistemic individuals do not form the basis of, but are rather an abstraction from, the community of inquirers. While the transcendental argument underwriting Davidson's position lacks power against Quine's naturalism, I argue that contemporary scientific philosophers have reason to take more seriously, first, a modest version, and second, an empirical variant inspired by Mead's theory.

5.1 Mead, Morris, and the Social Process of Semiosis

Charles Morris (1901–1979) is today primarily remembered for originating the distinction between syntax, semantics, and pragmatics. He earned his doctorate in 1925 at the University of Chicago, where six years later he took up an associate professorship. He traveled to Europe for his 1934 sabbatical and gave a paper at the Eighth International Congress of Philosophy in Prague, in which he waxed optimistic about future collaboration between pragmatists and logical positivists. Upon his return, Morris was instrumental in securing Carnap an appointment at the University of Chicago. Together with Otto Neurath, Carnap and Morris solicited and edited the monographs contributed to the *International Encyclopedia for Unified Science*, which was intended to be the signature publication of the Unity of Science movement. Yet despite these auspicious beginnings, Morris's subsequent philosophical career was not marked by great success. His major work in semiotics was poorly received, and his abiding preference for sympathetic synthesis rather than incisive criticism has made him easy for history to overlook.³

Although gatekeeping projects (such as articulating criteria by which to distinguish science from pseudo-science) were part of the Unity of Science movement, Morris's work emphasizes its inclusivity. He characterizes it in terms of a shared attitude he detects among American pragmatists, European logical positivists, and British empiricists alike, which he dubs "scientific empiricism":

> By this term is meant the temper which accepts propositions into the system of knowledge in proportion as they are verified by observation of the things or kind of things meant, but which does not want to exclude from consideration whatever rationalistic, cosmological, or pragmatic factors prove to be integral parts of the scientific method or edifice. (1937, 3–4; cf. 1938, 88)

Morris hopes that all inquirers who seek answers through their own (scientific) labor, rather than by looking backward to history or beyond to the divine, can find a home beneath this umbrella.⁴

Scholars have recently returned to the papers Morris published in the late 1930s in order to better understand Carnap's engagement with pragmatism.⁵ But to extract the criticisms regarding objectivity nestled within Morris's collegial and quixotic prose, I want first to consider the work he completed immediately prior. In the early 1930s, Morris was immersed in George Herbert Mead's lectures on social psychology. Mead had died unexpectedly in 1931, and despite being a leading pragmatist had written

---

³ Critics judged Morris's magnum opus, *Signs, Language, and Behavior*, narrowly behavioristic and insufficiently political – in the words of Margaret Schlauch, for instance, as "unsatisfactory to a Marxist scientist" (1947, 17) – in contrast to the alternative avowedly political semiotics being developed by those in the tradition of Ferdinand de Saussure, Claude Lévi-Strauss, and others. Morris's other philosophical interests proved unfashionable or positively objectionable, such as his use of William Sheldon's (1954) controversial research into correlations between body and personality types in an effort to articulate different ways of life "appropriate" to different peoples. For more on Morris's reception, see George Reisch (2005). (Reisch adds that Morris's writing style is unfortunately dense, "ever-polite and nonconfrontational" [44]).

⁴ The need to mobilize those friendly to science was particularly acute in Morris's home institution. In 1930, the University of Chicago's new president, George Hutchins, appointed the Thomist Mortimer Adler as professor of philosophy at an exorbitant salary, overriding the protests of the largely pragmatist department. Together with Richard McKeon, a specialist in Ancient Greek philosophy who championed historical textual analysis, Hutchins and Adler rejected both positivism and pragmatism and developed the Chicago curriculum around a humanistic Great Books program, with which both Morris and Carnap were out of step. Adler was an especially incendiary interlocutor, concluding his 1941 lecture "God and the Professors" with the following horrifically worded criticism of his colleagues: "Until the professors and their culture are liquidated, the resolution of modern problems will not begin [as] democracy has more to fear from the mentality of its teachers than from the nihilism of Hitler."

⁵ These are collected in Morris (1937). See Uebel (2013) and Mormann (2016).

no books. Keen to posthumously publish his mentor's ideas, Morris set about acquiring and editing copies of lecture notes from Mead's former students. The volume *Mind, Self, and Society: From the Standpoint of a Social Behaviorist* appeared in 1934, and was to become the most widely cited of Mead's works. Yet from the outset its fidelity was disputed.[6] Scholars continue to treat the text gingerly as a guide to Mead's views, since, as Hans Joas puts it, "[Morris] took such liberties in supplementation and emendation that one can never be sure whether a sentence is Mead's or Morris's" (1985, xii).[7] For that very reason, however, the book provides the careful reader with some insight into Morris's own ideas.

In Mead's view, human mindedness results only from the communal exchange of "significant gestures," vocalizations that have become meaningful (and meant) signs to their makers. "Until the rise of his self-consciousness in the process of social experience," he writes, "the individual experiences his body – its feelings and sensations – merely as an immediate part of his environment, not as his own" (1934, 172). He concludes that it is a mistake to conceptualize human experience thinly as populated with objects-as-stimuli that merely "lead [man] on and drive him away, as [they do] the well-mannered dog" (376). Rather, the objects that self-conscious individuals experience are socially conditioned. They exist within a field of possible interaction in the community, as what breaks when thrown, as what Father likes, as what makes Mother sick, and so on.[8] Human experience is populated with objects that are "collapsed acts," constituted by the range of things that can be done with, to, and because of them (1964, 134).

---

[6] Mead's colleague Ellsworth Faris, for instance, charges Morris with "[rearranging] the material in a fashion that will be deprecated by many who knew Mead and thought they understood him" (1936, 809).

[7] The most significant of Morris's interpretive glosses is characterizing Mead's position as "social behaviorism" at all, a moniker never used by Mead, which serves to bring his work in line with then current science, at the cost of obscuring the richness of his conception of behavior. See Cook (2013, 97ff) for an argument that Mead is better classified as a functionalist than a behaviorist, but cf. de Waal (2008, 147, 162). Christoph Limbeck-Lilienau conjectures that Morris was trying to draw a connection between Mead's work and Neurath's, whose 1932 "Sociology in the Framework of Physicalism" uses the phrase (2012, 10).

[8] Mead argues that humans possess a "role-taking" cognitive mechanism that allows us to align ourselves with others in our community and to learn about the nature of objects. (This mechanism also underwrites our ability to anticipate others' responses, and to intend our gestures to have a particular significance for them.) He imagines a child who grasps that its parent forbids something it finds desirable: "The child's capacity for being the other puts both of these characters [desirability and prohibition] of the object before him in their disparateness" (1934, 376). The child's grasp of what the object is – that is, how the object can enter into social action and be acted upon in various ways by different individuals – and, moreover, the object's *being* what it is, is the result of social processes.

In his introduction to *Mind, Self, and Society*, Morris draws out what he takes to be the significance of Mead's approach. For Mead, any "consciousness" theorists are tempted to ascribe to a pre-self-conscious individual is fundamentally distinct from "that which we term thinking or reflective intelligence" (1934, 165). Such an individual is not conscious *of* the objects it experiences. In Morris's words, for Mead, "the self, mind, 'consciousness of,' and the significant symbol are in a sense precipitated together" (1934, xxiii). And the radical consequence for empiricism is that epistemology must be grounded in the social arena, with a robust conception of the experience enjoyed by self-conscious inquirers.

Morris's articulation of Mead's account of scientific objectivity is especially striking: "On Mead's view the world of science is composed of that which is common to and true for various observers – the world of common or social experience as symbolically formulated" (xix). "The most important type of objectivity," Morris continues, is "social universality," which individuals approach to the extent that they share awareness of aspects of their experience:

> The individual transcends what is given to him alone when through communication he finds that his experience is shared by others ... the individual has, as it were, gotten outside of his limited world by taking the roles of others, being assured through communication empirically grounded and tested that in all these cases the world presents the same appearance. Where this is attained, experience is social, common, shared; it is only against this common world that the individual distinguishes his own private experience. (xxix)

What we term "objectivity" and "subjectivity" are thus poles of a prior, intersubjective continuum. At one extreme are those experiences shared by all. Moving through shrinking degrees of social plurality, we finally reach, at the other extreme, experiences we judge unique to ourselves. Empirical science, on this view, is the investigation of those features of the world that prove highly invariant, securing a high degree of social plurality or even true universality: "[Science] attains an independence of the particular perspective of the observer by finding that which is common to many, and ideally to all, observers" (xxix).

This conception of objectivity is irrevocably rooted in a society of inquirers. An objective truth is characterized as a truth for us all, not as a truth about an independent world. Objectivity is the view from everywhere, rather than nowhere. In turn, an individual's capacity for objective inquiry is explained through her membership in a community, not in terms of her relationship to the world. For the world of objects is a *shared* world of *social* objects, and the individuals who investigate it together only eventually cull perspectives that count as their own.⁹

Morris cheerfully adopts all of these views, describing his position even toward the end of his career as "near to that of Mead and the pragmatists" (1963, 87). His own work on semiotics tries to capture (not critique) Mead's insights within a broader account of semiosis, one capable of accounting for more primitive uses of signs studied by other psychologists.¹⁰ Such a synthetic ambition is, as I noted above, typical of Morris's approach, and also characterizes his interactions with the logical positivists he met in the mid to late 1930s.

Seeing the logical positivists as kindred spirits to the pragmatists – whom he goes so far as to call "bio-social positivists" – Morris hopes to inaugurate semiotics as an organon for unified science.¹¹ All sciences are sign-based human activities, he reasons, so semiotics could serve as a general framework to which scientists and scientifically minded philosophers working in different traditions could appeal in order to locate and understand each other's work.¹² He influentially distinguishes three aspects of semiotics. Syntax is the examination of the relation of signs to other signs. Semantics is the examination of the relations of signs to what they denote. Pragmatics is the examination of the relations of signs to their interpreters. Historically, Morris argues, formalists have attended to syntax, empiricists to semantics, and pragmatists to pragmatics. But crucially, it is the social process of semiosis which is fundamental, rather than any single aspect

---

⁹ This is not to rule out private experiences. As Morris observes, "Experience, of course, has its private aspects: one person's toothache is not another, any more than one person's perspective of a table is another's" (1937, 68). Nevertheless, he believes Mead's social behaviorism shows that "'My' has no meaning except over against 'your'," and that "unless there was a social or common dimension of experience the notion of private or individual experience would be without meaning" (35).

¹⁰ Morris replaces Mead's sharp binary between significant (i.e., human) and nonsignificant (i.e., nonhuman) signs with an account of how the process of semiosis plays out differently among groups of humans, nonhuman animals, and even solitary individuals. "The presence of two organisms at a happening which becomes symbolic to both," Morris writes, "is simply the occurrence of two conditioning processes which might just as well have taken place separately" (1925, 39). He hopes that viewing semiosis on a continuum eases a tension in Mead that he calls the problem of emergence: for how could we say anything about the nature of biological human experience – the stage from which we emerged – using symbolic language that only refers to *our* self-conscious experience (1937, 34–35)?

¹¹ See Morris (1937, 22) and (1938, 134). Cheryl Misak (2013) details how the pragmatists held various views congenial to logical positivism, such as opposing speculative metaphysics and defending a scientific, experimental attitude to social problems.

¹² In fact, since *all* human practices are mediated by signs, Morris hopes semiotics will ultimately foster greater mutual understanding between those pursuing scientific, artistic, religious, and political endeavors. "Signs serve other purposes than the acquisition of knowledge," he writes, "and descriptive semiotic is wider than the study of the language of science" (1938, 135).

of semiotics. The recognition of this point "provides the corrective to the one-sidedness of these attitudes when held in isolation" (1938, 74).

Morris diagnoses the protocol sentence dispute, for instance, as rooted in the Vienna Circle's lack of a "sufficiently general theory of meaning" (1937, 12, n.2). The word "meaning" is ambiguous, he insists, and "meanings are not to be located as existences at any place in the process of semiosis, but are to be characterized in terms of the process as a whole" (1938, 123). Meaning is thus not solely a semantic concept for Morris, unlike for many (including Carnap) who came after him.[13] As semiotics matures, Morris hopes, new terminology for our use of signs will be developed that will privilege semiosis in its entirety, allowing future inquirers to reconceptualize their activities in explicitly social terms.

Despite Morris's unflaggingly conciliatory tone in this period – "the temper of the age is ripe for synthesis," he writes, not division (1937, 6) – against the background of Mead's social behaviorism we may detect the lurking objection in scientific empiricism. For while it does no harm to focus temporarily upon a particular aspect of the process of semiosis (say, syntax), such an inquiry will inevitably use "terms which have all three dimensions and [employ] the results of the study of the other dimensions" (1938, 131). Semiotics is not a tree with three branches, affording a divide-and-conquer approach to be undertaken by specialists proceeding in relative isolation (having, perhaps, a passing interest in each other's work). Rather, syntax, semantics, and pragmatics are viewpoints that each provide partial insight into the social process of semiosis. Formalist, empiricist, and pragmatist investigations in turn *require* each other if they are to avoid a distorted view of the sign manipulations characterizing scientific practice.

Instead of developing this objection, Morris is content to display the virtues of pragmatist inquiry, gently admonishing that empiricists and formalists' "prejudice against the concept of social experience should not be allowed to go unquestioned" (1937, 68). The sole way to correct "the tendency to individualistic subjectivism" in historical forms of empiricism (which Morris diagnoses as "the accompaniment of an inadequate theory of mind") is "the discovery and elaboration of the social, objective, pragmatic aspect of mind" (1937, 67–68). Channeling Mead, Morris writes that "the human mind is inseparable from the functioning of signs – if indeed mentality is not to be identified with such functioning" (1938, 79). Although he

---

[13] Similarly, according to Morris, "'truth' as commonly used is a semiotical term and cannot be used in terms of any one dimension [that is, syntactic, semantic, or pragmatic] unless this usage is explicitly adopted" (1938, 119).

detects solipsism in Carnap's *Aufbau*, he praises Carnap's *Syntax* project as a clear (though conservative) characterization of the philosopher's role within scientific empiricism.¹⁴ Yet he thinks Carnap too quickly discards value judgments as meaningless. Philosophers *qua* language constructors have certainly provided important insights into "formal meaning, the place of rules in determining formal necessity … the foundations of mathematics, of physicalism, and the unity of science" (1937, 10). But since we do not merely mean formal meaning by "meaning," Morris argues, scientific empiricists should be able to study the empirical and pragmatic meaning of value terms.

As is well known, Carnap refuses Morris's invitation to empirical axiology and instead adopts an expressivist metaethics.¹⁵ But Carnap's proposals for securing scientific objectivity reveal a further divergence.¹⁶ In the *Aufbau*, he proposes viewing objects as purely structural definite descriptions constructed within an autopsychological constitution system. In making no essential reference to subjective experience, study of these formal objects will count as objective. By the *Syntax*, Carnap tolerates a variety of formal languages between which inquirers may freely choose given their current purposes. Having identified terms that will count as observational within a language selected for scientific inquiry, an inquirer may form and test objective hypotheses about the world.

---

¹⁴ Morris objects to the *Aufbau*'s reconstruction of intersubjectivity from intrasubjectivity, arguing that such a position "is forced to the conclusion that epistemologically the intersubjective world which science demands can only be 'inter' the subjects constructed in 'my' experience, so that intersubjectivity collapses into intrasubjectivity of a rather complicated sort" (1937, 32), and thus into solipsism. Yet, as Morris later observes, Carnap came to reformulate his "methodological solipsist" position in less objectionable terms, as the "trivial fact" that individuals need to observe others' observations via reports or testimony if they are to employ them (1963, 92). Nevertheless, Carnap's dismissal of the problem of other minds as a pseudo-problem, to be replaced by the question of whether or not to add psychological predicates to the reistic language (1963, 888), would plainly be inadequate to Morris, who accepts Mead's argument that such predicates (in application to a community) are required to account for the existence of self-conscious minds capable of inquiry.

¹⁵ According to Thomas Mormann (2007, 2010, 2016), this shows that the avowedly formalist Carnap took very little from his encounter with pragmatism. Yet Thomas Uebel (2010, 2013) reads Carnap as more sympathetic to pragmatist insights. Noting that Carnap incorporated Morris's semiotic terminology into his philosophy, Uebel attributes a bipartite metatheory to him under which there is simply a division of labor between formal studies of science, as represented in his own work, and empirical studies of science, as represented in Neurath's and the pragmatists'. Uebel judges Morris's semiotics inferior to this bipartite metatheory, since focusing upon the manipulation of linguistic scientific signs neglects the lessons that may be learned from other empirical sciences of science (2013, 543). Yet since Morris would view such other empirical sciences (e.g., the sociology of science) as themselves constituted by sign manipulations, they would certainly have had a role within his program. And one advantage of Morris's semiotics is its eschewal of a division of labor (which may, however unintentionally, lead to isolated parochialism), in favor of emphasizing unity and communication within a single overarching system.

¹⁶ As Uebel observes, Carnap does not conceive of himself as reconstructing the objectivity of knowledge, but as *constructing* a conception adequate to the aims of the unity of science (2001, 215).

From Morris's perspective, neither of these proposals for securing objectivity come near to Mead's social universality. A solitary inquirer ought not be entitled to call her investigations objective merely because she has abstracted from her own experiences or adopted a certain formalism. Seeking objectivity is rather a matter of seeking experiences shared among the community that is inquiring together.

Carnap justifies the pursuit of formal interests in isolation from pragmatics by subtly reworking Morris's semiotic categories. He reckons "descriptive semantics and syntax" – that is, the work of determining which language a group of inquirers are using – "part of pragmatics," but claims that "pure semantics and syntax … are independent of pragmatics" (1937, 13). Morris should have objected that this move elides the social nature of objectivity, which cannot be captured in syntactic or semantic terms alone. Carnap's tolerance about which language an inquirer chooses to use constitutes a permissive silence about the conditions under which a group selects a language for science, how its choice may be ratified and revoked, the number of inquirers who use such a language, and so on. From Morris's perspective, articulating objectivity demands the pursuit of both pure and descriptive *pragmatics*. Scientific investigations of our scientific practices (that is, descriptive pragmatics) will expose how they function and enable improvements, with the goal of achieving ever greater consensus about our shared experience of reality. And this work will be facilitated by the clarification and standardization of concepts appropriate for describing the users of signs (that is, pure pragmatics).

How might Carnap respond to this objection? The obvious target is Mead's social behaviorism, which might be judged a speculative exaggeration of the importance of other inquirers in furnishing an understanding of an individual's capacity for objective inquiry. Morris anticipates criticism that Mead's theory of mind is insufficiently positivistic:

> The extreme positivist – who is frequently the traditional epistemologist in disguise – would probably feel that in the strictest sense this position, while positivistic in tone, has insensibly moved beyond what can be empirically tested, and he may say, with a twinkle in his eye, beyond the limits of meaningful discourse. Is not role-taking, he goes on, only your (we will overlook his easy use of "your" and "mine") imaginative assumption of the attitudes and positions of others, as is evidenced by the facts of mistakes made in the process, so that you get outside of your experience only in imagination? (1937, 38)

His rejoinder is that we do not hypothesize or reconstruct interpersonal communication on the basis of individual experience. Human experience is rather the experience of successful communication, making the invocation of imagination a red herring. "The fact that no postulate is required as to the actuality of communication," Morris writes, "suggests the epistemological significance of a critical semiotic" (1937, 70). Enriching our account of the nature of human experience as arising out of the social process of semiosis entitles the positivist to build fellow inquirers into her account of the nature of observation and empirical testing. "I can observe that you expect what I do upon my use of words," he confidently proclaims, "and that the object which satisfies my expectations satisfies yours" (39–40).

However, equipped with the principle of tolerance, Carnap need not go on the offensive against Mead. He may instead observe that inquirers tempted by the pragmatists' social account of mind, experience, and objectivity are free to elaborate the process of semiosis in their own language. All Carnap asks of a language builder is that he "state his methods clearly, and give syntactical rules instead of philosophical arguments" (1934/1937, §17). The usefulness of employing a language for unifying scientific inquiry from an avowedly social perspective, as opposed to his own so-called individualist one, may be assessed at a later date.

If Morris is to undermine this maneuver, he requires an argument that Carnap's tolerance is itself biased. Notably, Carnap limits his plea for clarity in linguistic construction to syntax, not pragmatics – never demanding an account, for instance, of what will constitute adoption of the language. In his early work, Morris vaguely warns of the "unconscious self-deception" among certain formalists that lead them into producing "quasi-pragmatical" "pseudo thing-sentences" about semantic rules of usage – "rule" being, he contends, a term properly belonging to pragmatics – and insists that the "congenial attitudes evoked by [elaborate sign systems do not] constitute semantic rules" (1938, 119). But it is only with the benefit of hindsight three decades later that arguments about the results of Carnap's framing emerge, both in Morris's and Philipp Frank's contributions to Carnap's Schilpp volume.

According to Frank, Carnap's formalism has resulted in a philosophy of science powerless to combat the ideological forces that influence scientific practice (1963, 159). Without sufficient weight being given to the pragmatic aspect of semiotics, he argues, the choice of which theory to adopt in cases of underdetermination seems the free choice of each individual scientist, rather than being itself the proper object of scientific study and planning. In a similar vein, Morris observes that the attention of scientific philosophers has become lopsided toward mathematics and the physical sciences, doing a disservice to the "intensive studies in descriptive pragmatics now under way in psychology, the social sciences, and the philosophy of science" (1963, 90). Without the development and articulation of pure pragmatics, the human sciences have inevitably come to seem second rate and vague. Starting with syntax and not semiotics has yielded, according to Frank, a politically and culturally isolated philosophy of science, and to Morris, a distorted magnification of the physical sciences that misses the possibilities scientific empiricism opens up for the human sciences, and for scientific philosophy beyond the philosophy of science.[17] In his polite replies, Carnap concedes the interest of further research in pure and descriptive pragmatics, but is resolute about the enduring need to distinguish the internal, factual, truth-evaluable questions that may be framed within a precisely constructed language from the external, nonfactual, nontruth-evaluable decisions that may be made about languages (1963b, 861). For Carnap, it is only after adopting a constructed language that each inquirer can adequately define and thereby secure the objectivity of her investigations.

5.2 Morris's Scientific Empiricism and Quine's Naturalism

Like Morris, Quine may be read as perceiving an inadequacy in Carnap's neglect of pragmatics. In "Two Dogmas of Empiricism," Quine characterizes himself as espousing "a more thorough pragmatism" than Carnap in repudiating the analytic/synthetic distinction (1951e, 43). Although Quine later clarified that he did not intend to align himself with the American pragmatists with this remark (1991, 397), users of language – whose behavior Morris wanted to study as the pragmatic aspect of semiosis – are at the core of his critique. Quine argues that the analytic/synthetic distinction lacks any basis in scientific practice, and charges it with being a metaphysical intrusion in Carnap's otherwise scientific philosophy. Carnap is not convinced. In his view, we may define terms as we like when engaging in pure syntax and semantics, and he is proposing that inquirers adopt a formal language in which the analytic/synthetic distinction is specified in a certain way if doing so is useful. Languages currently in use only enter the picture when we turn to descriptive syntax and semantics.[18]

Unlike Quine, however, Morris holds on to the analytic/synthetic distinction, and sketches a theory of the "variable *a priori*" that comports with Carnap's *Syntax* project:

---

[17] Morris suggests that the scientific empiricist "might find that much of what has been called metaphysics can be explicated as the analytic portion of ... general or philosophical frameworks, a possibility that Carnap does not seem to have considered" (1963, 98). In this way, he believes, scientific philosophy "is compatible with the philosophical traditions of diverse cultures," since it becomes the "comparison, the criticism, and the proposal of the most general linguistic frameworks," as employed in different speech communities (97).

[18] See Ebbs (this volume) for an argument that Carnap misunderstood the force of Quine's attack.

> There is at any moment for thinking beings, an *a priori* in the sense of a set of meanings in terms of which empirical data are approached, and logical analysis may be regarded as following the structural lines of the *a priori* which support inference. This *a priori*, however, undergoes change through contact with the new data which are encountered through its use, and through changes in human interests and purposes. (1937, 51)

As Alan Richardson (2003, 13 ff.) has emphasized, Carnap would insist on a difference in kind between changes to our theory demanded by empirical investigation and changes to our language judged pragmatically useful, where Morris sees only our evolving theory. Because Morris denies Carnap's firm dividing line, Richardson concludes that he is pulled toward a Quinean holism. So, just how close are Morris's and Quine's views?

One connection is that according to both Quine and Morris, and in contrast to Carnap, certain metaphysical questions remain part of the scientific philosopher's bailiwick. Both judge ontological questions as meaningful and answerable. Quine tells us that "what distinguishes the ontological philosopher's concern [from the concerns of scientists] is only breadth of categories. [It] is scrutiny of … uncritical acceptance of the realm of physical objects itself, or of classes, etc., that devolves upon ontology" (1960, 275). Such a philosopher is "limning … the most general traits of reality" (161). Similarly, Morris defends what he calls "empirical cosmology," arguing that "[pragmatists] have all defended in one way or another an empirical equivalent to metaphysics in the form of a search for the generic features of all experience" (1937, 17).

Morris's example of an empirical cosmologist is A. N. Whitehead. At his best, Morris maintains, Whitehead strictly follows the method of "descriptive generalization," isolating features common to a set of experiences, and then experimentally generalizing to treat them as conditions for every experience.[19] The resulting theory is empirically testable, since counterexamples may be found. The goal of empirical cosmology, Morris writes, is "to erect a conceptual scheme of such generality that it is confirmed by all data. It differs from science in the narrower sense only in generality, and not in method nor in the security of results" (18). He judges

---

[19] To quell the logical positivists' doubts about the meaningfulness of such metaphysics, Morris sketches a fanciful thought experiment involving microscopic scientific empiricists living on a single human blood cell. Such microscopic inquirers, Morris claims, might – truly! – extrapolate from their experiences that they are living inside a macroscopic man, and yet, since they are unable to confirm that their empirically meaningful hypothesis is true, refuse to adopt the hypothesis into their theory of reality (1942, 220). His point is that so long as metaphysical claims are grounded in experience, positivists should regard them as meaningful.

Carnap wrong to think cosmology the province of science as opposed to philosophy, on the grounds that there is no clean break between the two:

> at the level of the widest system of knowledge the distinction between philosophy and science vanishes: a *unified completed science and an achieved philosophy would be identical.* But until that goal is reached it is possible to distinguish between conceptual systems progressively adequate to specific domains of experienced existence, and the attempts to formulate a system adequate to all domains whatsoever. This does not mean to dictate to science, but to use scientific results and the rich field of common life in the service of the most generalized science, that is, philosophy. (1937, 18–19; italics in original)

Much of Morris's position here is congenial to Quine. Like Morris, Quine sees no firm divide between philosophy and science, and defends the use of scientific results to inform problems that are traditionally called philosophical. However, in Quine's view questions of ontology are made intelligible against the background of a theory that has been regimented in the currently accepted canonical notation for science. First-order logic plus identity provides a crisp criterion for existence. Our scientific theory is our enlightened common sense, and the posits it demands are in what our reality consists. So unlike Morris's empirical cosmologist, the naturalized metaphysician does not seek to generalize from features of particular experiences, but is rather concerned with the formal regimentation of scientific theory.

In addition, the significance of testability distinguishes these defenses of metaphysics. For Morris, showing that Whitehead's theory is subject to empirical disconfirmation is necessary to certify it as empiricist. But for Quine, abandoning the analytic/synthetic distinction erodes the line between evidential and pragmatic considerations in theory choice. The appeals to parsimony and indispensability that characterize Quine's naturalized metaphysics are motivated in terms of what inquirers find useful as well as what they establish as true, a fact that allowed for a greater post-Quinean naturalist revival of speculative metaphysics than would have made Quine himself comfortable.[20]

Concerning objectivity, Quine is justly celebrated for integrating elements from both Neurath and Dewey (both of whom he credits explicitly) in his naturalism.[21] Neurath thought that the objectivity of a scientific

---

[20] Quine acknowledges in the opening paragraph of "Two Dogmas of Empiricism" that abandoning the analytic/synthetic distinction "[blurs] the supposed boundary between speculative metaphysics and natural science" (1951, 20). For an excellent discussion of the implications of this blurring, see Rosen (2014).

[21] For discussion, see Howard (2003).

theory was secured by its systematization and prediction of relevant protocol sentences (the statements which express elementary observations of an inquirer), but emphasized that all such theories are underdetermined by their evidential bases.[22] For Dewey, like Mead and Morris, the objectivity of a scientific theory is a function of its endorsement by a community of inquirers. Quine combines Neurath's underdetermination of theory with Dewey's focus on intersubjectivity in his account of observation sentences. (It is also worth noting what Quine does not take up from Neurath and Dewey: the role of social and political considerations in theory choice.[23])

Quine defines the stimulus meaning of a sentence for an inquirer as the ordered pair of stimulations that would prompt her to assent to and dissent from that sentence. An occasion sentence for an inquirer is one that requires stimulations to command her assent, in contrast to the standing sentences of her theory, to which she will assent in any situation. Finally, an inquirer's observation sentences are those occasion sentences that are independent of collateral information. For instance, assenting to "that is red" depends solely upon current stimulation in a way that "that is a bachelor" does not (maleness and lack of a spouse not being observable).

Quine next broadens this individualistic, behaviorist account of observation sentences to an inquiring community. Speakers whose sentences have similar stimulus meanings will count the same statements as observational. The scientific theory of a community is to be assessed in terms of how adequately it explains and predicts the intersubjectively defined observation sentences of the community, as well as other scientific values, such as overall simplicity, familiarity, etc. Quine takes each individual's theory to be objective because it is subject to evaluation at the communal scientific tribunal. Moreover, characterizing language as "our social art," he develops a theory of its acquisition that grounds our ability to refer in our interpersonal interactions (1960, ix). Ordinary objects, such as chairs and tables, no less than entities discovered by scientific inquiry, such as dark matter and genes, are theoretical posits. The origins of the ordinary objects we have posited may be "shrouded in prehistory," but nevertheless depend upon our shared inquiry (1960, 22).

---

[22] For more on the varieties of underdetermination, see Creath (this volume).

[23] Neurath sees room to pursue Marxist politics consistent with scientific objectivity in choosing to adopt whichever empirically adequate (and thus objectively true) theory best promotes socialism. Dewey defends a pragmatist agenda consistent with scientific objectivity in using results from the social sciences to actively cultivate the scientific attitude within society. In contrast, Quine's conception of science is effectively isolated from politics. More recent naturalists have sought to demonstrate the political import of scientific philosophy. Charles Mills (2007), for instance, presents his work in the philosophy of race as the result of expanding Quine's naturalism to the social sciences.

In building intersubjectivity into his account of observation sentences and adopting a conception of ontology as based in language that recalls Mead's claim that new objects are brought into existence as the result of social interaction, one might think that Quine successfully rehabilitates the epistemic significance of individual experience against Morris's concerns. Far from being overlooked, the community of inquirers now functions as the critical check against which individuals assess their theories, according to standards that are themselves open to investigation and criticism.[24] Yet Quine comes to believe that his early account of observation sentences is unsustainable. Once he recognizes, under pressure from Davidson, that there is no sense to be made of multiple inquirers "sharing," or even having "similar," stimulus meaning (since stimulus meanings are sets of stimulations and each individual has a unique neurology), he later describes inquirers sharing an *experience*, "jointly witnessing an occasion." They are capable of empathetically understanding each other's perspective as a result of having evolved parallel saliency standards through which they encounter their environment (1992a, 3).[25] Complicating Richardson's account of Morris's and Quine's relationship, then, we may wonder how far it is *Quine* who in his late work is ultimately pulled back toward Mead, Morris, and the pragmatist idea of social experience.

5.3 Davidson's Demand and Externalist Epistemology

Davidson's philosophy owes much to Quine, but he maintains that their views are importantly distinct even after Quine gives his new account of observation sentences. Davidson claims that his own "epistemology starts from intersubjectivity, that is, from the experience of sharing objectivity ... thinking presupposes intersubjectivity. This will remain an irreconcilable dispute between Quine and me" (quoted in Borradori 1994, 53–54). Although I have found no evidence of a direct connection between them, Davidson's position here clearly echoes Mead's.[26] Where Mead objects to the Cartesian philosophy that "set

---

[24] See Quine and Ullian (1978) for an account of the standards used to evaluate scientific theories.
[25] Commentators are divided over Quine's alteration. Anil Gupta, for example, calls socializing experience in this way "thoroughly non-Quinean," as it seemingly abandons the empiricist credentials afforded by focusing upon sensory stimulations (2006, 48, n.57).
[26] Indeed, Mead's approach anticipates Davidson's celebrated argument that we cannot make sense of an uninterpretable language: "I have been looking at language as a principle of social organization which has made the distinctively human society possible. Of course, if there are inhabitants on Mars, it is possible for us to enter into communication with them in as far as we can enter into social relations with them. If we can isolate the logical constants which are essential for any process

up a process of thought and a thinking substance that is the antecedent of [the] very processes within which thinking goes on," on the grounds that "thinking is nothing but the response of the individual to the attitude of the other in the wide social process in which both are involved" (1934, 260), Davidson similarly proposes turning Cartesian epistemology on its head:

> The empiricists have it exactly backwards, because they think that first one knows what's in his own mind, then, with luck, he finds out what is in the outside world, and, with even more luck, he finds out what is in somebody else's mind. I think differently. First we find out what is in somebody else's mind, and by then we have got all the rest. Of course, I really think that it all comes at the same time. (Quoted in Borradori 1994, 50)

Just as Morris detects an objectionable individualism in logical positivism, according to Davidson, "[Quine's] epistemology remains resolutely individualistic ... there is no reason in principle why we could not win an understanding of the world on our own" (2001b, 10) since "[Quine] makes the content of empirical knowledge depend on something that is not shared with others," (2001a, 291), namely, an individual's sensory stimulations.

Yet the force of this objection is unclear. Quine's naturalized epistemologist abandons Descartes's solitary methodological skepticism for a spot with the scientists crewing Neurath's boat (1969d, 16). And although he defines the empirical content of a particular statement as all those sentences of an individual's idiolect whose stimulus meanings are synonymous with the observation categoricals that the original statement implies, he generalizes by asserting that communities may associate the same *sentences* with the empirical content of a given statement (1992a, 16–17). Sentences, unlike stimulations, are social. Quine thus captures empirical content at the shared linguistic level, not at the isolated individual one, as Davidson claims. Observation sentences, while having a unique stimulus meaning for each speaker, are assessed interpersonally when they figure in scientific

---

of thinking, presumably those logical constants would put us in a position to carry on communication with the other community. ... The process of communication cannot be set up as something that exists by itself, or as a presupposition of the social process. On the contrary, the social process is presupposed in order to render thought and communication possible" (1934, 260). Like Davidson, that is to say, Mead judges all thinking bound by the same logical constants, which would be the entering wedge for any attempt to interpret the language of aliens with whom we could socially interact. Davidson goes beyond Mead in alleging that we are entitled to attribute language use and thought only to those others with whom we could (doubtless after a good deal of interpretive labor) communicate (Gupta 2006, 184). For Davidson, *our* social process is presupposed in order to render what *we* call thought and communication possible.

theory. This intersubjective requirement, Quine tells us, "is what makes science objective" (1992a, 5).

Nevertheless, Davidson maintains that awarding stimulations even this place in our theory of knowledge makes available a distorted conception of inquiry. The signature move of Quine's naturalized epistemology is to start in the midst of current scientific theory, according to which humans are physical objects equipped with sensory organs. In this context, Quine takes the epistemologist's task to be providing the "rationale of reification" (1990d, 3), an account of how knowledge of the world can arise from the sensory stimulations that humans, so understood, experience. Yet despite the communal tone much of Quine's rhetoric evokes, Davidson's point is that naturalists are now able to conceive of inquiry in solitary terms. My status as an inquirer can ultimately be traced, the naturalist may conclude, to the stimulations that resulted in my development of my theory. Naturalizing epistemology has blurred the sharp line distinguishing the objective thoughts individuals are capable of entertaining only once they engage in communal inquiry from the cognitive processes that may occur when nonsocial beings are stimulated by the environment.

Davidson's ruling of Quine's interest in stimulations as vestigial Cartesianism parallels the individualist hangover Morris finds in Carnap. Just as Morris hopes that acknowledging the irreducibly social aspects of self-conscious experience eliminates the need to treat belief in other minds as something dubitable, Davidson insists that the fundamentally intersubjective context in which the concept of objective truth is acquired vitiates the epistemological interest of individual sensation. It is only interpretation, he argues, that allows us to recognize a distinction between the way the world appears to us and a way that the world really is, a distinction that we record in our distinction between subjective belief and objective truth (1975, 170). And it is only by constructing an interpretive truth theory for others – and so attributing grasp of the concept of truth to them – that we recognize others as having minds (1973, 193). What remains once epistemology is externalized, Davidson argues, is a conception of inquiry as fundamentally social, in which objective truth does not reduce to intersubjective agreement on particular occasions, but functions instead as the goal of our shared endeavor.

Elsewhere, I have argued that Davidson is best understood as a *humanist*, in the sense that he treats the cluster of epistemic concepts that he analyzes, including truth, belief, thought, and justification, as firmly rooted in our conception of inquiry (Pearson 2011, 7–8). He holds, for instance, that it is a condition for being what *we* call a mind – for what *we* call having beliefs – that one be interpretable as having a mind. This humanist commitment is perhaps most explicit when he explains his motivation for externalism:

> [Aspects] of our interactions with others and the world are partially constitutive of what we mean and think. There cannot be said to be a proof of this claim. Its plausibility depends on a conviction which can seem either empirical or a priori; a conviction that this is a fact about what sort of creatures we are. Empirical if you think it just happens to be true of us that this is how we come to be able to speak and think about the world; a priori if you think, as I tend to, that this is part of what we mean when we talk of thinking and speaking. After all, the notions of speaking and thinking are ours. (Davidson 2001a, 294).

At the heart of Davidson's position is a transcendental argument about what we can indubitably know about ourselves and each other, given the nature of what we call inquiry. But this form of argument cuts no ice with Quine, whose naturalism rules out as supra-scientific any appeals to necessity. Even were all of the various claims Davidson makes about interpretation correct – such as that all beings we take to have minds are beings who have the concept of belief, who grasp the contrast between belief and truth, who are capable of being surprised, who have the capacity for error, and so forth – his judgment that the *only* way to acquire the contrast between truth and belief is through interpreting another mind amounts to *a priori* speculation. The naturalist may thus refuse to budge on the epistemic import of stimulation.

Putting aside Davidson's transcendental argument, however, there are two related ways of pressing the enduring need to socialize objectivity that I think scientific philosophers have reason to consider. The first is suggested by Mark Sacks's distinction between transcendental features and constraints. A transcendental constraint "indicates a dependence of empirical possibilities on a nonempirical structure, say, the structure of anything that can count as a mind" (2000, 213). Davidson took himself to have discovered transcendental constraints upon mindedness in his work on radical interpretation. In contrast Sacks writes that a transcendental feature "is significantly weaker":

> Transcendental features indicate the limitations implicitly determined by a range of available practices: a range comprising all those practices to which further alternatives cannot be made intelligible to those engaged in them ... those transcendental features of what we can currently envisage are not constraints on what is possible. (213)

This distinction opens room, consistent with naturalism, for a Davidsonian argument that conceiving of ourselves as investigating the world with other minds is a transcendental feature of the human practice of inquiry. If a Davidsonian can convince us that we cannot understand how anyone could grasp the contrast between subjective belief and objective truth without interpreting another inquirer – even if there is, unbeknownst to us, some such way – so that every form of inquiry we can make sense of is social, then we have no reason to think that attending to the experience of individuals will help us to better understand the nature of inquiry. The onus then falls upon the scientific philosopher to offer a credible alternative to Davidson's account of our grasp of truth.[27]

The second argument for socializing objectivity is suggested by Mead's theory of mind. Recall that Davidson saw two possible groundings of the conviction that social interaction partially constitutes thought: his own *a priori* one concerning the meaning of our concepts, and an empirical variant – "if you think it just happens to be true of us that this is how we come to be able to speak and think about the world" (2001a, 294). Mead's social behaviorism is just such an empirical theory. Where Davidson sought to persuade Quine of the need to socialize and externalize epistemology, of accepting that our theories presuppose the inquiring community, Mead develops an argument for socializing and externalizing metaphysics, of accepting further that the world we encounter is ours.[28] The objects of objective reality experienced by self-conscious minds are just as social as

---

[27] Lepore and Ludwig (2005) suggest that a solitary inquirer could triangulate with her past self and acquire the contrast between truth and belief. But the triangulation Davidson thinks necessary for the acquisition of this concept involves more than merely finding three things to occupy its vertices. We use the word "belief" to ascribe endorsed propositional contents *about* the world to minds. If the cognitive processes of the beings in the triangle are to count as beliefs, the triangle must somehow ground this relation of "aboutness." Davidson argues his interpersonal triangle does so because the occupants of the "I" and "you" vertices fix the parts of the world that their beliefs are about by mutually perceiving them. He makes the same point in another way by insisting that the reference of the concepts possessed by a nonlinguistic animal cannot be intelligibly located at any particular place in the casual chain from the world to its senses (1992, 119). Here again, it is useful to compare Mead's position that the experience of biological individuals is not *of* objects (1934, 165).

[28] Where Quine argues that members of a species evolve to have similar saliency standards for their environment, carving up their sensory experiences in parallel ways, a fact which means that our communication tends to proceed without difficulty when we project our own categories onto the experiences of our interlocutor, Davidson instead argues that members of a community count as talking about the same objects in virtue of triangulating with each other in acts of interpretation. The causal chains stemming from the distal object in the world to each speaker fixes the semantic content of our language. Mead turns this point around. Projecting oneself into the role of another is not only necessary for the emergence of significant signs in the social process, but also necessary for social objects themselves to exist. For Mead, objective reality is thus intersubjective reality. It is only in the process of attaining consciousness of the self through manipulating significant signs in a community that our reality comes into being.

the minds encountering them. Mead might have been brought to agree with Quine that scientific philosophy ought to begin *in mediis rebus*, but would deny that this starting point demands the rationale for reification. The reason is that, *contra* Quine, our scientific theories do not describe *us* as physical objects in a physical world. Psychology (i.e., social behaviorism) tells us we are not the biological individuals that would answer to such a description, but rather self-conscious ones that think about objects once we talk with others. Mead would thus applaud Quine's shift to discussing our joint witnessing of events and shared experience as evidence of maturation in his psychological views. Scientific philosophers should abandon individualistic fiction, and ground their theories within the inquiring community.

Coda: Obligations to Others

All the same, one might still wonder if this challenge to socialize objectivity amounts to a distinction that will make little practical difference to scientific philosophy. After all, Quine and Carnap both accept that science is a social activity, and wrangling over what should count as objectivity might seem merely terminological. If the worst that their so-called individualist abstraction of inquiry does is to allow inquirers to ignore or downplay the importance of their community, this seems a small price to pay for the fruits of their formal approach. But I think that we should beware this easy response. The way we conceptualize inquiry matters. As Peter Hylton has recently emphasized, naturalism – and, I would add, scientific philosophy more generally – is a position that comes with responsibilities as well as rights (2014a, 152). Too many philosophers eager for the prestige of such labels fail to live up to their demands, abrogating their responsibilities to other inquirers in their single-minded pursuits.[29] To take seriously the pleasing rhetoric of philosophers and scientists forming a community demands careful reflection upon the broader context of one's projects, centering rather than sidelining issues of social epistemology. The importance of this labor is only sharply in focus once we acknowledge that inquiry is a communal activity. To return to our point of departure, is an aspect of scientific empiricism that mattered deeply to Morris.

Although, together with Ernest Nagel, Quine and Morris were the philosophers most responsible for introducing logical positivism to America in

---

[29] In Pearson (2017), I argue that objectionable single-mindedness may also be detected in Quine's account of explication.

the 1930s, there is only one record of scholarly interaction between them.[30] It also involves Carnap. In 1941, both Carnap and Quine were among the respondents to Morris's paper "Empiricism, Religion, and Democracy." In this sweeping work, Morris elaborates how the scientific empiricist is empowered to offer sociocultural criticism by studying semiotics, and, indeed, must do so if crucial ground is not to be ceded to the scholastics:

> It is not enough that [the empiricist] limit his activity to the formulation and confirmation of scientific statements in the special fields of science. He must question the analysis of contemporary culture with which he is damned; he must attack the metaphysical super-structure which his opponents graft upon the edifice he so laboriously and cautiously erects; he must show that there is a way (or ways) of life – a rich, dynamic, satisfying life – compatible with his attitude; he must deny that his opponents have a monopoly on the defense of religious and cultural traditions of man; he must see to it that his own attitude clothes itself with esthetic, religious and political symbols adequate to serve in the enhancement and direction of life. (1942, 213)

Morris sketches how to give a semiotic analysis of a political term such as "democracy," or an entire text such as Hitler's *Mein Kampf*, making clear "the conditions under which the Nazi symbols have gained their power, the goals they express, and the essentially motivational character of their appeal" (232). Ultimately, he proposes that scientific empiricists construct new symbols with which to motivate and inspire the scientific attitude in culture, and which are capable of fulfilling the religious and political needs of humankind without bringing their users "into … emotional or doctrinal conflict with science" (236).

In their published responses, neither Quine nor Carnap are comfortable with Morris's choice to call the articulation of an empiricist value system a "religion." (Quine quips, "It has struck me before now that there are two cardinal methods, one favored by the conservatives in religion and the other by their more liberal brethren, for making the rest of us religious: conversion and definition" [238–239].) Yet it is significant that they are both

---

[30] In 1934, two years after Quine first visited Vienna, Carnap wrote to him about Morris's intention to visit the Circle and asked whether Quine knew him (Carnap to Quine, February 25, 1934). Quine did not answer this question in his response, but judging from a later remark of Morris's, the first direct contact between them was not until 1936, when Quine acceded to Morris's request for a letter in support of Carnap's appointment to Chicago (Morris to Quine, March 3, 1936). (Morris closes this letter, "Carnap has told me much about you and I hope before long we can meet.") Over the next few years, they wrote to each other with increasing frequency regarding arrangements for the Congress for the Unity of Science to be held at Harvard, but only met in person at the conference in 1939 (Morris to Quine, September 19, 1939). (Morris closes the letter, "It was nice to have come to know you personally.")

sympathetic to the project of promoting scientific values in our culture, and of clarifying our obligations to fellow inquirers. Whether or not scientific philosophers are ultimately persuaded to abandon the epistemic interest of individuals viewed in isolation from their community, reflecting on the challenge to socialize objectivity is a useful corrective for those of us whose tendency to engage in narrow scholarly pursuits may lead us to neglect our responsibilities to a broader culture struggling with, to name just one example, widespread distrust in epistemic authority.[31]

---

[31] For their useful feedback on earlier versions of this chapter, I would like to thank audiences at the Metropolitan State University at Denver and the seventh annual meeting of the Society for the Study of the History of Analytical Philosophy at McMaster University.

CHAPTER 6

Whose Dogmas of Empiricism?[1]

Lydia Patton

6.1 Conceptual and Thorough Pragmatism

A key development of early twentieth-century American philosophy is the exploration of pragmatist accounts of scientific language and behavior.[2] Two American schools of thought were developed, one in the 1920s, and another in the 40s. The first concentrated on developing an intensional account of meaning that could be a foundation for scientific and philosophical language and behavior. It was developed mainly by Clarence Irving Lewis, drawing on Charles Sanders Peirce, William James, and Josiah Royce, and he called it "conceptual pragmatism."

The second view questioned any but a hypothetical and in-practice distinction between intension and extension in scientific and philosophical languages.[3] This account was developed by Nelson Goodman, Morton White, and Willard van Orman Quine.[4] We can call it "thorough pragmatism," picking up on Quine's own language.

Quine describes his opposition to intensionalism about meaning as running deep and beginning very early in his career: "The distrust of mentalistic semantics that found expression in 'Two Dogmas' is ... detectable as far back as my senior year in college" (Quine 1991, 390). And Quine

---

[1] Clark Glymour mentioned to me that, in his view, Lewis was the real target of "Two Dogmas," which started me along this path. in October 2017, by the kind invitation of Sean Morris, I presented an early draft as a talk at the workshop "Carnap and Quine Reconsidered" in Denver. Commentary there from Morris, Gary Ebbs, Richard Creath, Sander Verhaegh, Peter Hylton, and James Pearson was instrumental in revising and refining the paper. A second version was presented at the Society for the Study of the History of Analytical Philosophy in 2018. During the question period, Sandra Lapointe, Thomas Ricketts, Warren Goldfarb, Gary Ebbs, and Landon Elkind contributed valuable comments and questions, from which the current version has profited immensely. I am sure I am forgetting some people who have discussed this project with me over the past years, for which I apologize.
[2] See Misak (2013) for the background and development of American Pragmatism.
[3] In White's case, sometimes drawing on John Dewey's reasoning in his Gifford lectures, *The Quest for Certainty*.
[4] White refers to "my fellow revolutionaries" Goodman and Quine in this context (1950, 317).

himself describes the problem at the root of "Two Dogmas of Empiricism" as driving a wedge between himself and his professors (later colleagues) at Harvard: "I was not abetted in my extensionalism by the Harvard professors of that time. Whitehead, C. I. Lewis, H. M. Sheffer, and E. V. Huntington all were soft on intensions and introspective meanings."

In the next breath, Quine remarks, "But a postdoctoral fellowship the next year took me to a kindred spirit in Czechoslovakia: the great Carnap" (Quine 1991, 391). Quine describes his disappointment at finding that Carnap was more committed to intensionalism than he'd thought at first. But reminiscences in "Two Dogmas in Retrospect" describe his opponents as Whitehead (whose introduction to *Principia Mathematica* appealed to intensionalism about propositional functions), Lewis, Sheffer, and Huntington (Quine 1991, 391). Carnap appears as a "great" philosopher and initial ally. The most negative remarks Quine makes about Carnap in "Two Dogmas in Retrospect" are about Carnap's positions on *modal logic* (1991, 392).

The account that follows will take the position that, in "Two Dogmas of Empiricism," one of Quine's deepest motivations was to draw a clear distinction between conceptual and thorough pragmatism, to defend thorough pragmatism, and to refute conceptual pragmatism. This motivation, I will argue, was much deeper than his desire to argue against Carnap.

There is a spate of recent scholarship on the relevance of pragmatism, and of his encounters with Lewis, to Quine's philosophy.[5] Still, Quine denied, repeatedly and in person, that he was influenced by "pragmatism."[6] It is entirely consistent with this statement that Quine's earliest philosophy at Harvard was developed partly *in opposition* to the "conceptual pragmatism" of C. I. Lewis.[7]

There is biographical evidence for such a claim. Morton White's autobiography mentions that Quine "had little love for these colleagues," which included Lewis (1999, 99). Quine's dislike of Lewis is well known. But

---

[5] See, among others, Sinclair (2012, 2016), O'Shea (2018), Mormann (2012a), Klein (2008), Järvilehto (2009), Franco (2020), Chang (2008), and Ben-Menahem (2016).

[6] Personal communications from Thomas Ricketts and Warren Goldfarb, during the question session at the Society for the Study of the History of Analytical Philosophy in 2018.

[7] I am grateful to Sean Morris for the following points: as early as the first 1934 Carnap lecture, Quine appeals to *Lewis's* definition of the analytic a priori, and to Lewis's account of the connection between them (Quine and Carnap 1990, 48). In the second lecture, where he does discuss Carnap, Quine is quite sympathetic to his view, and points out that on Carnap's account of logical syntax, "What is analytic for one language may not be analytic for another language. This is now seen from the formal definitions or explanations of the syntactic notion 'analytic'. But it is exactly the result which I came to last Thursday [in the first lecture] by an entirely different chain of reasoning" (78–9). Lewis is also discussed quite a bit in the White–Goodman–Quine triangular correspondence contained in White's autobiography.

there is also evidence to be gathered from analysis of the texts and arguments in and around "Two Dogmas," and I will marshal some of it in what follows.

In the account below, I will not argue that C. I. Lewis was a positive influence on Quine's reasoning in "Two Dogmas," but rather that the opposite claim is true: that Quine's longtime opposition to Lewis's positions finds its way, not only into the footnotes, but into the central positions of "Two Dogmas." Quine argued against Lewis because of Quine's deep commitment to extensionalism and opposition to intensionalism about meanings.

Nor will I argue that Quine didn't disagree with Carnap at all.[8] Instead, the account given below will begin from the recognition that there was no *deep* disagreement between Carnap and Quine on analyticity, but rather a difference of emphasis and method. Here, I draw on the work of others who have made similar points, especially Richardson (1997), Creath (2007), and Ebbs (2017).

The conclusion is that Quine's defense of "thorough pragmatism" against "conceptual pragmatism" is a fundamental motivation for his arguments against analyticity.

6.2 Carnap and Analyticity

Carnap and Quine did engage with each other on questions of analyticity, intension, and reduction over their long association, beginning around 1938.[9] However, many commentators, including Creath (2007), Richardson (1997), Sinclair (2016), and Ebbs (2017), have observed that the grounds of their disagreement are not what they may seem.

Creath notes that Quine's demand for behavioral criteria of analyticity seem ill-placed as a criticism of Carnap. The context for Carnap's views on the analytic/synthetic distinction was the development, over the nineteenth century, of nonclassical logics, non-Euclidean geometries, and relativity theory, which has the startling result that alternative geometries are possible as descriptions of the metric of physical space, as well as of the structure of perceptual space.[10] In both cases, there are no decisive experiments or principles. One of Carnap's initial motivations was to

---

[8] Frost-Arnold (2013) and Quine and Carnap (1990) catalogue Carnap's and Quine's early to mid-twentieth-century discussions.
[9] See the correspondence in Quine and Carnap (1990, 244–9), for their letters of early 1938 concerning intensional semantics. The discussion that followed covers decades.
[10] See Ebbs (2019) for a discussion of Carnap on analyticity.

develop models of theories which support univocal reference, perhaps via the "purely structural definite descriptions" of the *Aufbau*.[11] But Carnap realized, by the time of the *Logische Syntax*, that there was no fact of the matter that would decide between rival logical or geometrical systems.

> The standard candidates for a (Kantian) pure a priori – logic, arithmetic, and geometry – lost their status.[12] Carnap developed a way of dealing with this, supportinga distinction between claims that genuinely represent the world (the substantive claims) and others (the constitutive ones) that instead of representing give form, structure, and meaning to all the sentences of the language. The analytic sentences [...] are the ones whose acceptability is guaranteed by the constitutive principles. In turn the contradictory sentences are those whose unacceptability is guaranteed by the constitutive principles, and the synthetic sentences are all the rest. (Creath 2007, 326–7)

There is no bare fact of the matter that will decide between rival systems or frameworks. But there can be a distinction between statements that are depictions of the world, substantive claims, and the constraints on valid depictions. The latter are analytic: "in effect, the question of whether one can draw an analytic/synthetic distinction is exactly the same question as whether one can distinguish between the constitutive and the substantive" (Creath 2007, 327). Carnap argues that we must be able to "precisely delimit" any possible language for science, so that the constitutive and substantive claims can be distinguished (Richardson 1997, 157).

Quine's prima facie objection to Carnap's procedure is that there are no good behavioral criteria for the use of the word "analytic." Creath notes that

> Both before "Two Dogmas" and repeatedly after it Quine insists that he must be provided "behavioral criteria" for all intelligible terms and for "analytic" in particular. What Quine is demanding for analyticity, then, is essentially what Carnap demanded for physical length and other notions that were suitable for empirical science, namely, an indication in observational terms as to when the use of these notions was appropriate. (Creath 2007, 328)

But Quine wants more than that. He wants an account of what analyticity *is*, in behavioral terms. Carnap has specified only what the analytic statements of a given language are, but the point is to explain analyticity. As Quine writes,

> From the point of view of the problem of analyticity, the notion of an artificial language with semantical rules is a *feu follet*[13] *par excellence*. Semantical rules determining the analytic statements of an artificial language are of

---

[11] E.g., Richardson (1998, ch. 2; 2003, 176–177).
[12] Mormann (2012a, 2012b); O'Shea (2018); Stump (2015).
[13] Loosely, a red herring.

interest only in so far as we already understand the notion of analyticity; they are of no help in gaining this understanding. Appeal to hypothetical languages of an artificially simple kind could conceivably be useful in clarifying analyticity, if the mental or behavioral or cultural factors relevant to analyticity – whatever they may be – were somehow sketched into the simplified model. But a model which takes analyticity merely as in irreducible character is unlikely to throw light on the problem of explicating analyticity. (Quine 1951e, 34)

Reading "Two Dogmas," it stands out that Quine's fundamental attack is not on the analytic/synthetic distinction itself. Rather, it is on accounts of analyticity based on the intensional account of semantics, rather than on linguistic behavior. Quine wants a behavioral or operational definition of analyticity: how does a language user who employs language "analytically" behave differently from a language user who does so "synthetically"?

Carnap good-naturedly responds to Quine's requests in his 1955 "Meaning and Synonymy in Natural Languages." Carnap recognizes – as he well knew from his correspondence with Quine going back to the 1930s – that intensional meaning is the real issue for Quine, not analyticity per se. In the introductory sections of "Meaning and Synonymy," Carnap reflects on Quine's criticisms:

> Quine's criticism does not concern the formal correctness of the [intensional] definitions in pure semantics; rather, he doubts whether there are any clear and fruitful corresponding pragmatical concepts which could serve as explicanda. That is the reason why he demands that *these pragmatical concepts be shown to be scientifically legitimate by stating empirical, behavioristic criteria for them*. If I understand him correctly, he believes that, *without this pragmatical substructure, the semantical intension concepts, even if formally correct, are arbitrary and without purpose*. (1955, 34–35, emphasis added)

Carnap goes on to provide a clarification of "the pragmatical concept of intension in natural languages and to outline a behavioristic, operational procedure for it" (1955, 35).

First, Carnap characterizes the difference between his intensional and Quine's extensional semantics. For an intensional account of Carnap's kind, there must be some decisive consideration, some rule, about how to specify the intension of a term for which one knows the extension. For Quine's extensional account, once the extension of a term is specified, giving the intension is a "matter of choice" (1955, 37).

Carnap himself does not see the difference between Quine's view and his own as turning on the question of whether intensions are revisable (and thus whether analytic statements are revisable), or whether they are based on empirical generalizations or linguistic behavior. Carnap describes his own view as based on empirical generalizations, which, "like any other hypothesis in linguistics, can be tested by observations of language behavior" (1955, 37).

The process of determining an intension for Carnap is as follows:

> the determination of the intension of a predicate may start from some instances denoted by the predicate. The essential task is then to find out what variations of a given specimen in various respects (e.g., size, shape, color) are admitted within the range of the predicate. The intension of a predicate may be defined as its range. (1955, 39)

Users of natural language exhibit what Carnap calls *intensional vagueness*. That is, they do not explicitly determine the intension – the range of the predicate – ahead of time when employing a term. For instance, a language user, Karl, may use the term "human" frequently. But when Karl is asked whether Neanderthals are "human," he may not know the answer (1955, 40). This is "intensional vagueness": an inability of a language user to specify the precise range of a predicate.

Carnap observes that intensional vagueness is a particular problem in science. Characteristically, for him, intensional vagueness is a barrier to clear communication between speakers who disagree with each other about something. Without knowing the range of circumstances to which the terms and sentences they are using apply, there is no way to settle a disagreement between scientists, philosophers, or anyone at all.

Thus, Carnap recommends that scientific language be clarified so that users of scientific language can be sure of the intensional meaning of the terms they use. To Carnap, those who employ a term have in mind a range of ways the term can be applied. That range of application can be specified extensionally if we are only referring to syntax. But giving a semantic theory of an intensional language requires, for Carnap, an intensional metalanguage (Quine and Carnap 1990, 244–5).

Quine argues early on, in response to Carnap, that Carnapian tolerance goes too far in countenancing intensional meanings as one of the linguistic practices for which philosophy ought to account.[14] To Quine, one ought to "intolerantly" argue that intensional meaning is equivalent to metaphysics, which Carnap and Quine both were committed to overcoming (1990, 247–8). Quine effectively argues that, in tolerating the use of

---

[14] For the Principle of Tolerance and *Logical Syntax*, see Richardson (1994, 68–81); Ebbs (2017, §1.4). Both discuss the principle in relation to Quine.

intensional meaning in language, Carnap has allowed a form of metaphysics to enter by the back door. When Carnap employs artificial languages to discriminate between the sentences of a language held to be "analytic" and those held to be "synthetic," he covers with an artificial fig leaf the metaphysical practices of natural language users.

It seems Quine is missing the point of the principle of tolerance. The principle is not intended to discriminate between meaningless and meaningful languages in terms of what it will *account for*. Rather, tolerance states that you may set up whatever logic you like, and whatever form of language you like. There are two ways that tolerance helps to cut against metaphysics, in Carnap's system. First, "Carnap proposes that we avoid the vocabulary and methods of metaphysics by using only those language systems S such that for every sentence s of S, all investigators agree on how to evaluate s" (Ebbs 2017, 22). Second, "Metaphysics is handled … through retranslation from the material mode into the formal mode of speech. In this way, various claims that look to be about mathematical objects are shown to be disguised descriptions of languages for mathematics or proposals to adopt such languages" (Richardson 1994, 75). The same is true of intensional meaning in natural language. Claims that look to be about "essences" or "synonymy" turn out to be formal claims about how to evaluate the range of predicates, for instance (Carnap 1955).

Intensional meaning is used in natural and scientific language, and Carnap is obliged by the principle of tolerance to account for that use. He does so by developing an intensional metalanguage to account for the semantics associated with intensional meaning. But that metalanguage consists only of rules for the use of language, and the method of replacing essential claims with rules is intended to *dissolve* metaphysical claims about the reality or being of intensions, essences, or relations of synonymy. The entire point of Carnap's strategy is to distinguish reasoning in the metalanguage, about how to evaluate claims using intensions, from metaphysical claims about essence (1950a, 318).

Hence, as Creath (2007) and Richardson (1997) conclude, Carnap and Quine appear to be "talking past each other" (Richardson 1997, 152). Ebbs (2017) argues that they are pursuing very similar agendas. Both are, in a sense, "working from within," Carnap beginning with informal mathematics and natural languages and setting up frameworks to clarify them, and Quine beginning with the results of science and accounting for how they are achieved (Richardson 1994, 81).[15] Nonetheless, there is a clear difference of method between the two.

---

[15] See Verhaegh (2018) *Working from Within* for a full account.

For Quine, one begins with a general empiricist commitment to investigating the justificatory status of all claims on the basis of experience. The analytic nature of logic and mathematics is then brought in to show how they in fact do best on the question of confirmation by experience. Thus, the verificationist connection of meaning and confirmation and the intimate relation between meaning and truth for analytic claims show how such claims are confirmed come what may. For Carnap, however, we begin with a range of possible languages for science. Each of these languages must be precisely delimited in order to be considered as a genuine language at all. This delimitation induces an analytic/synthetic distinction for each language. (Richardson 1997, 157)

Despite these differences, a puzzle arises for reading "Two Dogmas" in the context above.

Once we have clarified the differences in method between Quine and Carnap, the challenge Quine mounts on analyticity in "Two Dogmas" doesn't seem to be truly aimed *at Carnap*. First, Quine has misunderstood Carnap's project if he thinks Carnap is in the business of giving an *empirical* account of analyticity. "Given that Carnap explicitly denies that his own attributions of analyticity are intended to have empirical content, it is illegitimate for Quine to demand that Carnap produce behavioral or other empirical criteria that will show that these attributions do have such content" (Creath 2007, 329).

Ultimately, Quine does seem to understand Carnap's project. Quine – and his fellow "revolutionary" Morton White – are perfectly aware that Carnap is aiming at an account of analyticity for artificial languages, that is, for "precisely delimited" languages (Richardson 1997) that are aimed at clarifying the meaning claims within natural or scientific languages in practice. Quine and White both say rather derisory things about Carnap's project here, arguing that Carnap is giving too bloodless an account of analyticity to really capture it. White comes right out and admits that he and Quine are doing something quite different from what Carnap is doing:

> But these [artificial] languages are the creatures of formal fancy; they are dreamed up by a logician. If I ask: "Is 'All men are rational animals' analytic in L_1?" I am rightly told to look up the rule-book of language L_1. But natural languages have no rule-books and the question of whether a given statement is analytic in them is much more difficult. We know that dictionaries are not very helpful on this matter. What some philosophers do is to pretend that natural languages are really quite like these artificial languages. (1950a, 321)

White is clearly talking about Carnap. But Carnap was not trying to *explain* natural language as a form of behavior. He was trying to *explicate* the forms within natural and scientific languages[16] – *if any such forms exist* – insofar as they can be made to support claims that the truth conditions of any sentence within that language can be evaluated by any competent speaker of the language.

Carnap's point was that we should be tolerant: we should give speakers the benefit of the doubt when they are laying down their languages and choosing the rules of those languages. If someone wants to import metaphysical claims about essences, Carnap's strategy is to give them enough rope to hang themselves.

Quine and White argue, instead, that we should begin with intolerance: that intensional semantics should be ruled out of any philosophical account of (natural) language, because the varied forms of linguistic behavior that fit under the term "analytic" have no essential connection with each other. Their disagreement begins with Carnap's strategy. They don't think Carnap should go even as far as he does to countenance intensional meanings (this is exactly how Quine describes his relationship to Carnap in 1991).

Moreover, White and Quine don't think that the argument that natural and scientific language users actually make valid claims about intensional meaning can be made to stick. In other words, they think Carnap is wrong to say that analytic and synthetic sentences can be distinguished from each other. Because of the principle of tolerance, Carnap argues that we must respect the assertions of language users that they do make such a distinction. And he says that if we consider the question as making sense internal to a given framework or language, we can make sense of attributions of "analyticity" to particular sentences.[17]

Quine's objections to intensional accounts are not limited to Carnap, and they did not begin with his response to Carnap. Quine says that "the great Carnap" initially seemed like an ally in his campaign. This makes sense, given that Carnap – in his own words! – was committed to accounting for intensional meaning only as much as an entomologist is committed to investigating fleas and lice (Quine and Carnap 1990, 245). Nonetheless, Quine thinks that Carnap should be even more intolerant. After all, Carnap is intolerant of Heideggerian metaphysics, and Quine puts intensional meaning on the same chopping block.

---

[16] Mentioning both here is not a claim that these are essentially distinct for Carnap: he explicitly says they are not.
[17] Here, we are treating that distinction as an "internal" question, in Carnap's sense in "Empiricism, Semantics, and Ontology."

But attacking Carnap can't be Quine's real motivation for going after intensional meaning. It can't even be Carnap who Quine is accusing of reifying intensional meaning in the first place. First, Carnap explicitly denies that he does so, and for reasons that are entirely consistent with Carnap's overall approach: he considers intension only because of tolerance. And Quine takes Carnap at his word about that, from their earliest correspondence onward (1990, 244–8 and throughout). Second, Carnap's method of accounting for intensional meaning is to introduce artificial language rules that *clarify* intensional meanings, but do not reify them.

Who, then, is the culprit? Who has turned intensional meaning into essence and wedded it to the word? Carnap is this person's unwitting henchman, but he is not the villain. So who is?

The villain lurking behind "Two Dogmas of Empiricism" is Clarence Irving Lewis. This is hardly a surprise. Reportedly, Quine had personal and professional reasons to resent Lewis.[18] Moreover, as Sinclair (2016, 2012) and Murphey (2012) detail, Quine was engaged deeply with Lewis's philosophy at Harvard from early on in his graduate career, taking Lewis's course on "the theory of knowledge, in which *Mind and the World-Order* was the main text. Among Quine's manuscripts there are three papers that he wrote for that class" (Murphey 2012, 6).

Sinclair argues that elements of Lewis's view are behind Quine's approach in "Two Dogmas" and elsewhere. This is clearly the case, but I believe that the influence of Lewis on Quine was in large part negative. Quine defined his own view in opposition to Lewis's view, for two reasons.

First, Lewis recognized that meanings and essences were closely tied, and he argues against essences as distinct from sense-objects. But Lewis argued that we must countenance fixed conceptual categories and schemas in order to clarify intensional meanings, because only then can we give a proper account of the truth of analytic claims. Lewis was willing to go as far as a kind of "Platonism" to support this view. Carnap never said anything approaching this. But a view of this kind is clearly Quine's real target in "Two Dogmas."

Secondly, Lewis tied his account of essences to a Peircean approach to empirical investigation in language and in science. Lewis's conceptual schemas involve holding the intensional meanings of, and the connections between, certain concepts fixed in order to interrogate changing phenomena. For Lewis, the meanings of concepts must be held fixed "come what will" (in Lewis's words) for us to be able to investigate and evaluate nature using language. Again, Carnap did not take this to be true. Carnap's

---

[18] White (1950).

account at this time was that our conceptual and linguistic frameworks can be varied at will (Carnap 1950a).

Two of Quine's most famous pronouncements in "Two Dogmas" are anticipated in Lewis:

> Meanings may be considered without reference to any applications they may have to existent things; but to reify them as another kind of objects than the sense-presented, is uncalled for and may lead to the ancient fallacy of ascribing to essences some kind of cosmic efficacy. (Lewis 1946, ix)
>
> Meaning is what essence becomes when it is di)vorced from the object of reference and wedded to the word. (Quine 1951e, 22)
>
> Since [the a priori] is a truth about our own interpretative attitude, it imposes no limitation upon the future possibilities of experience; that is a priori which we can maintain in the face of all experience, come what will. (Lewis 1929, 231–2)[19]
>
> Furthermore it becomes folly to seek a boundary between synthetic statements, which hold contingently on experience, and analytic statements which hold come what may. Any statement can be held true come what may, if we make drastic enough adjustments elsewhere in the system. (Quine 1951e, 40)

In the first case, Quine and Lewis are reasonably in agreement; in the second, Quine is clearly taking a position in opposition to Lewis. In the section following, we will see why.

6.3 From Conceptual to Thorough Pragmatism

Clarence Irving Lewis's account of the a priori embodies a precise tension between a linguistic account of meaning, and a pragmatic account of how meaning is linked to the interpretation and anticipation of experience. Here, we will focus on a specific problem Lewis himself identifies in *An Analysis of Knowledge and Valuation* of 1946. First, Lewis emphasizes the importance of theories of meaning to "contemporary empiricism." Meaning, Lewis says, has replaced metaphysics as the central pursuit connected with epistemology. We can "certify" a priori knowledge "by reference to meanings alone" (1946, ix).

Lewis considers two possible accounts of "meaning" in *An Analysis of Knowledge and Valuation*: the type that "has no dependence on language" and is unrevisable, and the type that is operational, pragmatic, or behavioral in character. The second type argues that

---

[19] As we'll discuss in the section following, the real source of the second view is much earlier: Peirce's "method of tenacity" in "The Fixation of Belief" (1877).

sign-function, or mediating function, which attaches to the given content in perception and marks it as cognitive, depends simply on the question whether or not the empirical eventualities which are signalized actually ensue when the mode of action is adopted. *Meaning in this sense is anticipation of further experience, associated with the presentational content; and the veracity of it concerns only the verifiability or non-verifiability of expected consequences of action.* (Lewis 1946, 15, emphasis added)

Quine could endorse this account. But Lewis's operational, pragmatist account of meaning always coexists, in Lewis's view, with another account, in which meanings are fixed, unalterable, and independent of language use. Lewis argues for an intensional theory of meaning (1946, 37), and he argues that intensions must be held fixed for analytic truth to be possible.

But Lewis does not argue that that the fixation of meaning is merely conventional. His response to the syntactic-linguistic theory of intension is to argue that there is something about analytic truth, and something about intension, that cannot be captured linguistically.

However much linguistic symbols are subject to convention and arbitrary rules, this freedom to stipulate and manipulate does not extend to the meanings which are symbolized. An analytic statement says something, and something whose factuality is independent even though it is not existential in significance (1946, ix).

An analytic truth expresses some content. That content may be expressed via symbols that are given significance by convention. But the content of an analytic truth, the "factuality" of it, is distinct from the syntax of the language in which it's expressed.[20] Analytic truth and intensional meaning are closely linked.

To Lewis, we have access to analytic, a priori truths by examining our own conceptual schemes and interpretations: "it becomes unnecessary to suppose that a priori truth describes some metaphysically significant character of reality ... What we know independently of sense-particulars, we can assure by understanding our own meanings and the connections of them with one another" (Lewis 1946, ix). Lewis concludes with a remarkable anticipation of Quine's remarks on meaning in "Two Dogmas" (1951e, 22): "Meanings may be considered without reference to any applications they may have to existent things; but to reify them as another kind of objects than the sense-presented, is uncalled for and may lead to the ancient fallacy of ascribing to essences some kind of cosmic efficacy" (Lewis 1946, ix). Meaning is "independent of sense-particulars" and

---

[20] It is possible that Carnap would have agreed with this statement, but for entirely different reasons.

independent of linguistic convention, but it is not for that reason part of a different realm of being than the sensible world. Instead, it consists of the definitions, interpretive frameworks, and "conceptual schemas" that we use to interrogate that world.

Thus, one might expect Lewis to summarize his analysis of meaning by repeating his warnings about endowing meaning or essence with "cosmic efficacy." Instead, he makes a startling statement: "Once the intensions of language are fixed, one can no more affect these meanings or alter their relations than one can alter the facts of existence by talking about them in a different dialect. On that point Platonic realism is nearer to the truth" (Lewis 1946, ix). Two years later, Lewis, Quine, and Morton White would all be at Harvard, and Quine and White (with Goodman in correspondence, at Penn) would engage in discussions about overthrowing intensional meaning.[21] Lewis was among the targets of their criticism, and certainly his account of analytic truth would have been a natural rallying point.[22] The idea that analytic truth, while not grounded in a realm of being distinct from the empirical, is nonetheless fixed and inalterable, would have been a point of disagreement. Moreover, what does Lewis mean by appealing to "Platonic realism"?

Lewis's point of reference would have been Raphael Demos, Lewis's colleague at Harvard during his entire career (1920–53). Demos's 1939 *The Philosophy of Plato*[23] lays out a theory of the a priori as part of Plato's theory of truth and knowledge. According to Demos, "Plato resorts to an ontological hypothesis, maintaining that unless we posit a realm of forms, the distinction between knowledge and opinion will not stand" (1939, 175). Knowledge of truth requires "communion" between "exact and distinct ideas" and a world that reflects them. The empirical world alone cannot support knowledge of the truth, because

> The empirical world evades numerical determination; its qualities lack purity. Thus, it fails to provide objects for exact and distinct ideas. And yet, since clear and definite knowledge exists, there must be a real world, beyond that of experience, which renders such knowledge possible. This world consists of the forms, each of which is pure, definite, and precisely what it is. Hence our ideas can be precise and exact and distinct. And as forms are in communion with each other, universal and necessary knowledge is possible. (1939, 175–6)

---

[21] White (1999, 99–104); White (1950, 316–8), *passim*; Creath (1990a, 35).
[22] White (1950, 323); White (1999, 102–3).
[23] Demos's *The Philosophy of Plato* (1939) came out ten years after Lewis's *Mind and the World-Order* (1929). I am arguing for an in-person influence. Lewis and Demos were colleagues beginning in 1920. Demos's book draws on a number of (much) earlier papers.

On Demos's reading, the Platonic forms explain the order and regularity of the phenomena without appeal to essences. They are *hypotheses* – Plato uses this word – that can be used in explanations.

In *Mind and the World-Order* (1929), Lewis explores many of the same questions Demos does in *The Philosophy of Plato*: knowledge, truth, and the a priori. Both investigate the conditions for finding order in nature, and argue that some hypothesis regarding ideas or conceptual schemas is necessary to explain and interpret that order. To Lewis, concepts "must be determined in advance of the particular experience to which they apply in order that what is given may have meaning. Until the criteria of our interpretation have been fixed, no experience could be the sign of anything or even answer any question" (1929, 230).

Lewis was concerned with the conditions of *truth* in his theory of the analytic and the a priori. His account requires that a priori truth be analytic, and be arrived at through "explication or elaboration of the concept itself" (1929, 231). He goes on to describe the a priori as "definitive or explicative, representing principles of order and criteria of the real."

> Since it is a truth about our own interpretative attitude, it imposes no limitation upon the future possibilities of experience; that is a priori which we can maintain in the face of all experience, come what will … The a priori is knowable simply through the reflective and critical formulation of our own principles of classification and interpretation. (1929, 231–2)

Lewis argues that certain a priori elements must be held firm as a priori, analytic interpretive categories to be able to interrogate experience and come up with any coherent answers. If we do not hold any such categories firm, Lewis argues, nothing in experience can reliably designate anything else, and phenomena dissolve into uninformative disorder.[24]

The similarity between the emphasized passage in Lewis and Quine's remark should be clear: "Any statement can be held true come what may, if we make drastic enough adjustments elsewhere in the system" (1951e, 40). But the superficial similarity hides a deep difference.

Quine's statement is that a person may cling tenaciously to a *belief that a claim is true*, in the face of increasingly recalcitrant evidence.[25] Here Quine acknowledges not only Lewis, but also Charles Sanders Peirce, who had elaborated "the method of tenacity" in "The Fixation of Belief." The

---

[24] This is, of course, one of the Kantian elements in Lewis's system, apparent throughout *Mind and the World-Order*.
[25] Quine's statement – and Peirce's method of tenacity – anticipate Leon Festinger's concept of cognitive dissonance.

method of tenacity consists of "taking as answer to a question any we may fancy, and constantly reiterating it to ourselves, dwelling on all which may conduce to that belief, and learning to turn with contempt and hatred from anything that might disturb it" (1877, §5). Peirce warns the reader vividly about the method's drawbacks:

> [M]en cling spasmodically to the views they already take. The man feels that, if he only holds to his belief without wavering, it will be entirely satisfactory. Nor can it be denied that a steady and immovable faith yields great peace of mind. It may, indeed, give rise to inconveniences, as if a man should resolutely continue to believe that fire would not burn him, or that he would be eternally damned if he received his ingesta otherwise than through a stomach-pump. But then the man who adopts this method will not allow that its inconveniences are greater than its advantages. (1877, §5)

Lewis does not seem to have been moved by Peirce's warnings. But Lewis thought the a priori, and intensional meaning, are *definitional* and *interpretive* only. It is neither a priori nor analytic that (A) one should receive all one's food through a stomach-pump. It *is* at least analytic, for Lewis, that (B) food nourishes human beings, and one cannot understand A without B.

Quine's statement undermines Lewis's attempt to preserve a definitional, interpretive a priori analytic within the pragmatist framework. Quine sees Lewis's analytic or "definitional" statements and categories as on a par with empirical beliefs about stomach-pumps and fire. There is nothing special about intensional meaning, synonymy, or "definitional" or "interpretive" categories that should allow us to cling to them tenaciously. As White (1950) notes, an "essential predication" such as "man is a rational animal" is just as questionable from a Quinean point of view as any empirical statement that can be compared with experience. Quine applies Peirce's criticisms of the "method of tenacity" to any philosopher who clings to fixed interpretive categories, despite the lack of evidence that they differ essentially from empirical claims.

Lewis and Quine agree on the dictum of another Pragmatist father, William James, that the criterion of meaning is anticipation of future experience.[26] Lewis argued (1949) that we settle meanings of terms a priori through analytic definition, and then figure out how terms refer by pragmatic means. This is influenced by the pragmatism of James, who was Lewis's professor at Harvard.[27] The passages from *Analysis of Meaning* that give operational definitions of meaning, as opposed to fixed ones, are very Jamesian. The idea is

---

[26] See Hookway (2008) for the relationship between Quine, Peirce, and Lewis.
[27] For the influence of James on Lewis and Quine, see Ben-Menahem (2016).

that we can employ terms to specify intensions conventionally, and then the cognitive significance of those very terms will be revealed in their significative function in representation, again in experience.[28]

From Quine's perspective, though, Lewis did not accept the full import of James's research, and of research in the Jamesian tradition generally. Using Peirce's criticisms of the "method of tenacity" against him, Quine argues against Peirce's "pragmatic maxim": that the meaning of a term or concept is the difference it makes to future experience. Peirce argues that holding certain concepts fixed allows us to control, and to investigate, what difference they make to our experience or to our interpretation of experience.

Quine argues that holding certain beliefs or categories, or Lewis's "conceptual schemas," fixed in order to interpret experience is impossible in practice. Quine knew perfectly well – and stated explicitly – that you can *try* to make certain statements "analytic" by throwing yourself under the bus for them. But that is not the same as arguing that certain statements have a privileged status as unrevisable, even with respect to another domain of statements.

It is possible in practice, Quine says, to try to hold certain statements as fixed, and to argue that they define the meaning of statements one will evaluate in the future. It might seem that this strategy would have good holist credentials. But, Quine says, it is precisely when we are holists about meaning that this strategy fails:

> It may seem that Peirce's maxim can be reconciled with the holistic view in this way: we can construe Peirce as meaning that the meaning of a statement consists in the difference which the truth or falsity of that statement would make in future experience supposing all other statements to be held fast in point of truth value. But ... [i]f you vary the truth value of one statement, presumably you should concomitantly vary the truth values of other statements which are logically implied by the given statement, or which logically imply it. Also presumably you will want to vary other statements which are connected with the given one by so-called well-established empirical laws.[29]

Thus, Quine rejects Lewis's attempt to rescue a conceptual pragmatist version of the analytic a priori. Lewis's definition of meaning as anticipation of experience (1946) coexists, in Lewis's work, with another account, according to which the intensions of terms are fixed. Lewis argues that only if the intensions are held fixed can we interpret our experience in the first place.

---

[28] See Klein (2008) for an account of James's conventionalism.
[29] 1951c, transcribed in an Appendix to Verhaegh (2018).

For Quine, the *only* criterion of meaning is anticipation of future experience. It is true that one can, in practice, try to hold fixed a certain statement whose terms are intended to *define* the character of future experience. That would be Quine's version of Peirce's maxim that "the meaning of a statement consists in the difference which the truth or falsity of that statement would make in future experience." If that meaning is taken as analytic, it is held fixed: the difference it makes in future experience is held to be unrevisable. But, Quine objects, that is not possible on a holist view. If the meaning of every sentence, especially every logical sentence, makes a difference to the meaning of every other, then it is not possible in practice to hold one sentence fixed with respect to all other sentences.

6.4 Conclusion

Lewis's account embodies a tension between:

1. his operationalist account of meaning, according to which the meaning of a term is in the difference it makes to the anticipation of experience; and
2. his desire to rescue analytic truth, which leads him to argue that the intensions of terms should be held fixed.

Lewis's "conceptual pragmatism" involves using stable intensional meanings as tools to investigate nature and to establish truth.

The argument against the analytic/synthetic distinction in "Two Dogmas of Empiricism" is largely aimed at rooting out Lewis's conceptual pragmatism, which Quine finds to be a species of "intensionalism." Quine's "more thorough pragmatism" sees no scope for holding meanings fixed, and argues for holism and extensionalism. First, as White argues, one of the hidden goals of the criticism of the analytic/synthetic dualism was "to show that if analysis is a fundamental task of philosophy and if the result of an analysis is expressed in an analytic statement, then a philosophical statement is removed only in degree from a scientific generalization" (1999, 103). There is no distinction in principle between inductive generalizations and intensional definitions or meanings.

As a result, Quine maintains that there is no difference in principle between analytic and synthetic statements: "my reservations over analyticity are the same as ever, and they concern the tracing of any demarcation, even a vague and approximate one, across the domain of sentences in general" (Quine 1991, 396). Here is where Carnap enters the picture.

Quine does not find Carnap's strategy of defining analyticity relative to a given language, perhaps by introducing artificial rules, to be persuasive.

However, on my reading, Carnap is collateral damage. While he was not sympathetic, Quine did not have a deep objection specifically to Carnap's attempts to introduce linguistic methods that "precisely delimit" (Richardson 1997) analytic from synthetic statements in a given language.[30] Instead, Quine was opposed, more generally, to *any* attempts to distinguish analytic from synthetic statements grounded on intensional accounts of meaning.

When we try to find the deep disagreement between Carnap and Quine, as Creath, Richardson, and Ebbs have observed, we strike out. But C. I. Lewis held a view that Quine opposed strongly. According to Lewis, analytic truth – which is worth preserving – can be rescued only if we hold intensional meanings fixed. Lewis's method was aimed at saving a Platonic realm of truths about the world, the content of which is distinct in principle from "scientific generalizations." Lewis's method, commitments, and conclusions are the underlying target of "Two Dogmas."

---

[30] From early on, Quine is quite careful to indicate where he is discussing Carnap, and where he is not. This is true even in the Carnap lectures, where Quine uses Lewis's definition of the analytic a priori, not Carnap's (Quine and Carnap 1990, 48). And it is true in "Carnap and Logical Truth," where the discussion of Carnap himself comes rather late in the paper. Thus, it is not legitimate to infer that Quine is discussing Carnap wherever he criticizes "intensional meaning" or "analyticity." Quine is usually quite explicit throughout his career when he is discussing Carnap. I am grateful to Sean Morris for making this point.

PART III
Carnap and Quine on Logic, Language, and Translation

CHAPTER 7

Reading Quine's Claim that Carnap's Term "Semantical Rule" Is Meaningless

Gary Ebbs

In §4 of "Two Dogmas of Empiricism" (henceforth "TD4"), Quine criticizes Carnap's methods of defining analyticity in terms of "semantical rules." His criticisms end with the provocative claim that Carnap's term "semantical rule" is meaningless. Many careful and informed readers, including Carnap himself, are unconvinced by what they take to be Quine's reasons for making this claim. The consensus in the literature about TD4 is that Quine's criticisms fail to show, by Carnap's standards, that Carnap's definitions of analyticity in terms of "semantical rules" are unsuccessful.[1]

Quine did not share this view. In being critical of Carnap's definitions of analyticity in terms of "semantical rules," Quine thought that he was applying Carnap's own standards more carefully than Carnap did. As Quine later put it, he "was just being more carnapian than Carnap" (Quine 1994b, 154).

I now think Quine was right about this. Readers fail to see the force of Quine's criticisms in TD4 for two main reasons. First, Quine does not explain in TD4 how his formulation of the problem of analyticity engages with Carnap's work. Second, Quine does not say enough about Carnap's methods and commitments to make it clear to his readers that there is no adequate carnapian response to his criticisms. Properly viewed, however, as I shall argue, Quine's criticisms reveal that Carnap conflates two senses of "semantical rule," the first of which, though clear, is of no use in defining analyticity, and the second of which, though integral to Carnap's method of defining analyticity, is unexplained by Carnap's definitions.

---

[1] E.g., Bonhert (1963); Carnap (1952/1990); Creath (2004); Ebbs (1997, ch. 5); Friedman (2010, 673–678); George (2000); Martin (1952); and Richardson (1998, 220).

7.1 A Puzzle about Quine's Formulation of the Problem of Analyticity

Quine launches his criticisms in TD4 by asserting that

(Q) The problem of analyticity is to make sense of '*S* is analytic for *L*' for variable '*S*' and '*L*'. (1951f, 33)

Immediately after asserting (Q), he states his central conclusion in TD4, namely, that

(C) The problem of making sense of '*S* is analytic for *L*' for variable '*S*' and '*L*' retains its stubbornness even if we limit the range of the variable '*L*' to artificial languages. (1951f, 33)

Quine's arguments for (C) begin with the following remark: "For artificial languages and semantical rules we look naturally to the writings of Carnap. His semantical rules take various forms, and to make my point I shall have to distinguish certain of the forms" (1951f, 33). Three pages later, after presenting his arguments for (C), he writes, "Not all the explanations of analyticity known to Carnap and his readers have been covered explicitly in the above considerations, but the extension to other forms is not hard to see" (1951f, 36).

These two remarks, which frame Quine's arguments on pages 33 through 36 for (C), respectively, make clear that in these pages Quine is criticizing Carnap's explanations of analyticity.

In his major works in semantics (Carnap 1939, Carnap 1942, and Carnap 1947), Carnap proposes definitions of analyticity for the most widely accepted and clearest sorts of artificial languages, namely, those with a first-order logical grammar (Martin 1952, 43). As Quine notes in §3 of "Two Dogmas of Empiricism" (30), each such language is equivalent, given an appropriate choice of definitions, to what I shall call a basic first-order language, that is, a first-order language without identity, names, or function terms.[2] If Carnap's methods for defining analyticity for basic first-order languages are unsuccessful, then his methods will also be unsuccessful when applied to more complex types of artificial languages. I shall therefore read Quine's criticisms in TD4 as directed against Carnap's methods of defining analyticity for artificial basic first-order languages, and reformulate (Q) as (Q'):

(Q') The problem of analyticity is to make sense of '*S* is analytic for *L*' for variable '*S*' and '*L*', where the range of the variable '*L*' is limited to artificial basic first-order languages.

---

[2] See also Quine 1960, §§37–38, §47; Quine 1986b, 63–64; Boolos, Burgess, and Jeffrey 2002, §19.4.

Many readers of TD4 are puzzled by (Q) and any of its qualified yet general forms, such as (Q'). Carnap expresses the chief source of this widespread puzzlement as follows: "In case Quine's remarks are meant as a demand to be given one definition applicable to all [language] systems, then such a demand is manifestly unreasonable; it is certainly neither fulfilled nor fulfillable for semantic and syntactic concepts, as Quine knows" (Carnap 1952/1990, 430). In support of this complaint Carnap cites Martin, who writes, "at present we have no truth concept of the immense generality Quine demands of the concept of analytic" (Martin 1952, 44). In the same paper Martin also alludes to some of the technical considerations that lead to Tarski's undefinability theorem, according to which no consistent language L rich enough to express elementary arithmetic contains its own "true-in-L" predicate.[3] This theorem implies that there is no single Tarski-style definition of "true-in-L" for basic first-order languages L. Since carnapian definitions of "analytic-in-L" imply that a sentence S is "analytic-in-L" only if S is "true-in-L",[4] the undefinability theorem also implies there is no single carnapian definition of "analytic-in-L" for variable "L." It would therefore be unreasonable for Quine to demand a single definition of analyticity that applies to all languages. The puzzle is that it is unclear how else to read (Q), or any qualified yet general form of it, such as (Q').

7.2 The Puzzle Solved

Quine knew Tarski's undefinability theorem and it is absurd to suppose that he forgot or overlooked it when he wrote TD4.[5] To solve the puzzle one must therefore find an interpretation of (Q) that is compatible with the undefinability theorem. I shall construct such an interpretation by highlighting overlooked aspects of Quine's summary and clarification, in Quine 1953b, of Tarski's method of defining "true-in-L" for a basic first-order language L.[6] In this summary, Quine emphasizes that applications of the pattern

(T) "_____" is true-in-L if and only if _____

endow "true-in-L" "with every bit as much clarity, in any particular application, as is enjoyed by the particular expressions of L to which we apply [it]" (Quine 1953b, 138).[7]

---

[3] Tarski first sketched the undefinability theorem in Tarski 1936. See Enderton 1972, 228–229, for a lucid proof of it.
[4] Carnap 1939, 13; Carnap 1942, 64; Carnap 1947, 10; Schilpp 1963, 901.
[5] Quine 1940, ch. 7, applies Tarski's undefinability theorem to prove that arithmetic is incomplete.
[6] Quine (1953b) is Quine's reply to Martin (1952). See Creath (1991, 362–363).
[7] For similar formulations this tarskian insight, see Quine (1940, 4); Carnap (1942, 26); Carnap (1946, §3); and Carnap (1963a, 60).

The clarity of the pattern (T) does not depend on any meaning we associate with the string of letters t-r-u-e that occurs in the defined term "true-in-L." The latter is a one-place predicate with no internal logical structure. Without any loss of significance, we may replace the pattern (T) by (K):

(K) "_____" is K if and only if _____.

The stipulation that "Jones smokes" is true-in-L if and only if Jones smokes, for instance, is just as clear to us as the stipulation that "Jones smokes" is K if and only if Jones smokes.

Particular instances of the pattern (T) or (K) are "partial definitions of truth" (Tarski 1944, 344). To develop a complete definition of "true-in-L" that avoids the liar paradox, Tarski argues, one needs to distinguish between an *object language L* that does not contain any predicate coextensive with the predicate "true-in-L" that occurs in applications of (T) to sentences of L, on the one hand, and a *metalanguage ML* in which we define "true-in-L" for L, on the other. He shows how to construct consistent definitions of "true-in-L" that imply all and only the "partial definitions" of "true-in-L" that can be obtained by applying (T) to particular sentences of L.[8] Quine emphasizes that "In Tarski's technical construction ... we have an explicit general routine for defining truth-in-L for individual languages L which conform to a certain standard pattern and are well specified in point of vocabulary" (Quine 1953b, 138). For most basic first-order languages, Tarski's method of defining "true-in-L" requires that we first define "satisfies-in-L" for the basic predicates and the logical constants of L. If the metalanguage *ML* contains the object language L, then definitions in *ML* of "satisfies-in-L" for the basic predicates of L may be specified disquotationally. We may stipulate, for instance, that

x satisfies-in-L "smokes" if and only if x smokes.

More generally, applications of the disquotational pattern

(Sat) x satisfies-in-L "_____" if and only if x _____

endow "satisfies-in-L" with as much clarity any one-place predicate of L to which we apply it. Similar disquotational patterns for defining "satisfies-in-L" can be formulated for predicates of L that have two or more argument places. Once definitions of "satisfies-in-L" for the basic predicates

---

[8] Kripke 1975 describes circumstances in which Tarski's method of constructing definitions of "true-in-L" does not help us to avoid paradox. I shall nevertheless assume that there are consistent applications of Tarski-style definitions of "true-in-L" that suit Carnap's and Quine's purposes, which are different from Kripke's.

of L are given, one also can define "satisfies-in-L" for sentences containing truth-functional connectives and for sentences containing quantifiers. One can then stipulate that a closed sentence S of L is "true-in-L" if and only if every sequence of objects satisfies-in-L S.[9]

When our metalanguage ML contains an object language L, the identity relation on sentences is in effect a translation relation between L and ML. Biconditionals of the form (T) therefore satisfy Tarski's Convention T, according to which an adequate definition of "true-in-L" entails all and only biconditionals of the form

X is true-in-L if and only if p

where "X" is a name of a sentence S of L, and p is a translation of S (Tarski 1936, 187–188).

This observation takes on a special significance in the context of Quine's investigation of Carnap's methods of defining semantical terms such as "true-in-L" and "analytic-in-L." Let me digress now to explain why. There are four key points.

First, Carnap distinguishes between pure semantics and descriptive semantics. In pure semantics one constructs artificial languages and investigates their logical properties without conducting, or relying on the results of, any investigations of the semantical properties of "historically given" languages, that is, languages that are already in use, such as English or a technical refinement of English that is used in physics, mathematics, or another special science (Carnap 1942, 12–13). In pure semantics, "We lay down definitions for certain concepts, usually in the form of rules, and study the analytic consequences of the definitions. In choosing the rules we are entirely free" (Carnap 1942, 13). In descriptive semantics, by contrast, one investigates the semantical properties of historically given languages (Carnap 1942, 11). Descriptive semantics is an empirical science that includes methods for observing and describing how speakers of an historically given language use its expressions, and is therefore part of what Carnap calls pragmatics, which studies how particular speakers use linguistic expressions (Carnap 1942, 9 and 13). In descriptive semantics we formulate and choose semantical rules with the goal of making sense of how the speakers of an historically given language use its linguistic expressions. Such choices, although not always uniquely determined by the pragmatical facts (Carnap 1939, 6–7), are never entirely free.

---

[9] See Tarski 1936; Quine 1986b, ch. 3; and Ebbs 2009, §2.8.

Second, according to Carnap, it is only in pure semantics, in which our choice of rules is entirely free, that "true-in-*L*" and "analytic-in-*L*" may receive exact definitions (Carnap 1939, §§4–7; Carnap 1942, §14; Carnap 1952/1990, 427; Carnap 1963d, 918), and the significance and clarity of the pure semantical terms are independent of any pragmatical facts about how sentences of historically given languages are used.

Third, the notion of translation that is relevant to Tarski's Convention T is not restricted to the sorts of stipulated formal translation relations discussed above, but also encompasses translation relations between an historically given object language *L* and a metalanguage that does not contain *L*. Translation relations of the second sort can only be established on the basis of empirical observations about how expressions of the relevant historically given languages are used.

Fourth, Quine was aware of these three points.[10] It is therefore crucial both to Carnap's efforts to define "analytic-in-*L*" in pure semantics and to a proper understanding of Quine's formulations of his criticisms of those efforts, that by exploiting the identity relation on expressions, we may construct exact definitions of "true-in-*L*" that satisfy Convention T without relying on methods or concepts of descriptive semantics.[11]

In accord with this conclusion, Quine's summary of Tarski's method of defining "true-in-*L*" in Quine 1953b supports the following two observations:

(I) For any given basic first-order language *L*, applications of the patterns (T) and (Sat) in a metalanguage that contains *L* are as clear to us as the sentences and predicates of *L* to which we apply the patterns. Our understanding of the significance of the applications does not depend on any prior meaning we associate with the expressions "satisfies-in-*L*" or "true-in-*L*."

(II) Tarski provides an explicit general routine for defining "true-in-*L*" in terms of "satisfies-in-*L*" for basic first-order logical languages. The routine can be used to generate definitions of "satisfies-in-*L*" and "true-in-*L*" of the sort described in (I).

Thus, for any given artificial basic first-order language *L*, we can use Tarski's method to define "satisfies-in-*L*" and hence also "true-in-*L*", and

---

[10] The Quine–Carnap correspondence in the 1940s shows that Quine knew Carnap's writings very well. See Quine and Carnap 1990.

[11] In pure semantics one may also stipulate how the expressions of an object language *L* are to be translated by expressions of our metalanguage *ML* and define "satisfies-in-*L*" so that an application of "satisfies-in-*L*" to an expression *e* of *L* is as clear to us as conditions we express by using the expressions of *ML* that, according to our stipulations, translate *e*.

thereby make sense of "*S* is true-in-*L*" for every sentence *S* of *L*, in pure semantics, even though, by Tarski's undefinability theorem, there is no single definition of "true-in-*L*" for all artificial basic first-order languages.

What does it mean, however, to say that Tarski-style definitions of "satisfies-in-*L*" and "true-in-*L*" are "semantical"? Tarski notes that "*Semantics* is a discipline which, speaking loosely, *deals with certain relations between expressions of a language and the objects* (or "states of affairs") *"referred to" by those expressions*" (Tarski 1944, 345; Tarski's italics). His methods enable us to construct precise new "semantical" relations between objects and expressions of a given formalized language *L*, namely, "satisfies-in-*L*" and "true-in-*L*", the latter of which we may use to affirm an infinitude of sentences *L* (Tarski 1944, 359). It is in this sense of "semantical" that Tarski's method of defining "true-in-*L*" exemplifies a clear and successful strategy for making sense of a semantical idiom of the form "*S* is φ-in-*L*" for variables "*S*" and "*L*."

We should therefore read (Q), Quine's formulation of the problem of analyticity, not as requiring that there be a single definition of "analytic-in-*L*" for variable "*L*", but as requiring that there be a method analogous to Tarski's for defining "analytic-in-*L*" for variable "*L*." This solves the puzzle of §7.1.

7.3 Comparing "True-in-*L*" with "Analytic-in-*L*"

A crucial question for both Carnap and Quine is whether the method Tarski developed for defining "true-in-*L*" for variable "*L*" can be extended to yield a method for defining "analytic-in-*L*" for variable "*L*." Quine 1953b is doubtful:

> See how unfavorably the notion of analyticity-in-*L*, characteristic of the theory of meaning, compares with that of truth-in-*L*. For the former we have no clue comparable in value to (T). Nor have we any systematic routine for constructing definitions of 'analyticity-in-*L*', even for the various individual choices of *L*; definitions of 'analyticity-in-*L*' for each *L* has seemed rather to be a project unto itself. The most evident principle of unification, linking analyticity-in-*L* for one choice of *L* with analyticity-in-*L* for another choice of *L*, is the joint use of the syllables 'analytic'. (Quine 1953b, 138, with "(7)" changed to "(T)")

Here Quine suggests that to solve the problem of analyticity we would need a method of defining "analytic-in-*L*" that satisfies conditions analogous to conditions (I) and (II) above, namely:

(I') For any given basic first-order language *L*, there is a schema applications of which yield definitions of the form "*S* is 'analytic-in-*L*' if and only if _____", where both what fills the blanks in the schema and the significance of the resulting definitions are specified independently of any prior meaning we associate with the expression 'analytic-in-L'.

(II') There is an explicit general routine for constructing, for each basic first-order language L, a definition of 'analytic-in-L' that satisfies condition (I').

To satisfy these conditions, Quine thinks, we would need a clue "comparable in value" to (T) that enables us to specify a "systematic routine for constructing definitions of 'analyticity-in-L'." Citing TD4, 32–36, Quine 1953b (138) claims we have no idea how to satisfy either of these conditions, and hence no idea how to solve the problem of analyticity, which, as he reminds us earlier in the paper, "was recognized as the problem of construing 'analytic for L' for variable 'L'" (Quine 1953b, 134). He thereby invites us to read his criticisms in TD4 as efforts to show that we have no idea how to satisfy (I') and (II'), and hence no idea how to solve the problem of analyticity, as formulated by (Q').

7.4 Summary of §§7.1–7.3

Observations (I) and (II) of §2 solve the puzzle of §7.1 and suggest that the only way to solve the problem of analyticity would be to construct definitions of "analytic-in-L" that satisfy conditions (I') and (II') of §7.3. Quine 1953b invites us to read his criticisms in TD4 as efforts to show that we have no clue about how to satisfy either condition (I') or (II').

7.5 Carnap's Methods of Defining "Analytic-in-L"

Carnap's methods of defining "analytic-in-L" are guided by the assumption that just as a Tarski-style definition of "true-in-L" must satisfy Tarski's Convention T, so an adequate definition of "analytic-in-L", or "A-true", must satisfy what I will call Convention AT (where "AT" is short for "analytically true"):

> Convention AT: A sentence S is **A-true** in language L if and only if S is true in L in such a way that its truth can be established on the basis of the semantic rules of L alone, without any reference to (extra-linguistic) facts. (Paraphrase of Carnap 1947, 10; see also Carnap 1942, §14; Carnap 1939, 13)

Carnap assumes that both the narrowly logical truths of a language L and truths that follow logically from "meaning relations" between predicates of L may be defined as "analytic-in-L" in accord with Convention AT.

Carnap and Quine agree that the narrowly logical truths of a language L can be specified semantically. It is this notion of logical truth that Quine loosely sketches in TD: "If we suppose the prior inventory of *logical* particles, comprising 'no', 'un-', 'not', 'if', 'then', 'and', etc., then in general a logical truth is a statement which is true and remains true under all reinterpretations of its components other than the logical particles" (1951f, 22–23). Quine 1940 presents a precise version of this definition, which Carnap endorsed in discussions with Quine in 1941 (Frost-Arnold 2013, 155–156). Carnap's preferred version of the definition is as follows:

> We call a true sentence logically true, in the narrow sense, when its truth is already established by the meanings of its logical constants. As a technical term, (explication) for this concept we take 'L-true'.
>
> D1. A sentence S is called L-true (in the language system LS), if it is either a true sentence without descriptive constants or results from one by substituting in descriptive constants. We say that S_1 L-implies S_2 or that S_2 follows logically from S_1 when $S_1 \supset S_2$ is L-true. (Carnap 1958a, 81)

Carnap assumes that Tarski-style definitions of satisfaction for sentences of LS that contain logical constants imply that the L-true sentences of a language system LS are "true in virtue of semantical rules, without any reference to (extra-linguistic) facts." Quine does not.[12]

Carnap sought to extend this strategy to construct definitions of a wider class of truths based on "meaning relations" between predicates. He once suggested (Carnap 1947, 4–15) that a Tarski-style definition of "true-in-L" for a given language L determines which sentences of L follow logically from meaning relations between predicates of L. As Quine pointed out to him (Letter 110 (1943) in Quine and Carnap 1990, 357; and Quine 1951f, 36), however, a definition of "true-in-L" does not clarify the relations between the meanings of the basic predicates of L.

Carnap's final, preferred method of defining A-true-in-L is to include a set of "meaning postulates," or A-postulates, among the "semantical rules" of L.[13] To a first-order language that contains the basic predicates "is a bachelor" and "is married", for instance, one may decide to add either "$\forall x((x$ is a bachelor$) \supset (x$ is married$))$" or "$\forall x((x$ a bachelor$) \supset \sim (x$ is married$))$" as a meaning postulate. The second choice fits better with pre-established uses of the words "bachelor" and "married." But such considerations belong to descriptive semantics, not pure semantics. In pure semantics we are free to

---
[12] Quine 1960, 65, note 3.
[13] Versions of this idea can be already found in Carnap 1934/1937, Carnap 1939, and Carnap 1942.

decide which "meaning postulates" and other "semantical rules" to adopt for a new artificial language (Carnap 1952, 67–68).

To select a set of sentences of L to be its meaning postulates is to place constraints on the meanings we may assign to the basic predicates of L that occur in the postulates. To assign meanings to the basic predicates is to construct a definition of "satisfies-in-L" for them that, together with definitions of "satisfies-in-L" for the other basic expressions of L, enables us to define "true-in-L" (Carnap 1939, 10).[14] If we aim to clarify the pretheoretical notion of analyticity and we assume, with Carnap, that a sentence of L is analytic only if it is true-in-L, we will choose to construct a definition of satisfaction that implies that the sentences we have selected to be the "meaning postulates" of L are true-in-L. To do so, we must pair "meaning postulates" of L with sentences that we already independently accept as true in our metalanguage. For this reason, as Carnap himself observes, although we are free to construct the logical syntax of an artificially constructed language in any way we like, if our goal is to interpret the inference rules of the stipulated logical syntax so that they preserve truth-in-L, we are not equally free when we interpret it (Carnap 1939, §12; Ebbs 2011, §§1–6).

Carnap, nevertheless, thinks we can clarify the idea that "meaning postulates" are not only true-in-L, but true-in-L "independent of extralinguistic facts," by showing (a) we can deduce from our definition of true-in-L that the meaning postulates are true-in-L without the addition of any other premises, and (b) in our deduction we only rely on the so-called "semantical rules" of L. Carnap's preferred definition of A-true-in-L, which is designed to satisfy (a) and (b), runs as follows:

> Let A be the conjunction of the A-postulates of a given language, then for this language we lay down the following definition.
> D2. A sentence S is called A-true when A L-implies S. (Carnap 1958a, 82)

When "L-true" and "A-true" are defined following the patterns D1 and D2, respectively, "all L-true sentences are A-true, although not all A-true sentences are L-true" (Carnap 1958a, 81–82).

Carnap's strategy for defining analyticity has three additional features that are important to understanding Quine's criticisms in TD4. First, D2

---

[14] Carnap writes, "My proposals for the explication of analyticity have always been given for a formalized (codified, constructed) language L, i.e., a language for which explicit semantical rules are specified that lead to the concept of truth [as defined by Tarski's methods]" (Carnap 1963d, 918).

is structurally similar to carnapian specifications of truths of a systematized natural science. Carnap recommends that we systemize our physics, for instance, in an artificial basic first-order language L by (first) laying down a set P of sentences of L that we call "P-postulates", and (second) adopting the following definition:

D3. A sentence S is called P-true when P L-implies S.[15]

Carnap assumes that we will define the set of P-postulates of L so that the P-truths of L are not A-true. We do so, in part, by stipulating that "A" and "P" denote disjoint sets of sentences of L, the sets of A-postulates and P-postulates of L, respectively. But Carnap thinks the difference in significance of the two types of definitions is not explained solely by such stipulations. In addition, he supposes that the A-postulates, unlike P-postulates, are stipulated to be not only true-in-L, but true-in-L in a special way, independent of "extra-linguistic facts."

Second, the concepts of pure semantics, such as the concept of an A-postulate, do not include "prescriptive components" (Carnap 1963c, 923). "The so-called rules are meant only as partial conditions of a definition" (Carnap 1963c, 923). Carnap uses the word "rule" in semantics, he explains, "only to conform to the customary usage in logic." Thus the distinction between "A-postulates" and "P-postulates" is not that one set of postulates is prescriptive, or normative, and the other is not.

Third, according to Carnap, neither the "A-postulates" nor the "P-postulates" of a language L can be revised without ceasing to use language L (Carnap 1963d, 921). The crucial distinction between the two is that the "A-postulates" are among the "semantical rules" that supposedly "determine the meanings" of L's linguistic expressions.

7.6 Quine's Criticisms in TD4 (Part One)

We are now positioned to reconstruct Quine's criticisms in TD4 of Carnap's use of the term "semantical rule." They begin as follows: "Let us suppose, to begin with, an artificial language L_0 whose semantical rules have the form explicitly of a specification, by recursion or otherwise, of all the analytic statements of L_0. The rules tell us that such and such statements, and only those, are the analytic statements of L_0" (1951f, 33). For example, suppose that L_0 results from language L when certain of L's sentences are selected to be "A-postulates" of L_0. Then L_0 is just like L except

---

[15] Carnap 1934/1937, §51; Carnap 1939, §23; and Carnap 1958b, §§42 and 48.

for also including these "A-postulates" among its "semantical rules." Let A be the conjunction of all of L_0's A-postulates, and define:

D2'. A sentence S of L_0 is **analytic for L_0** if and only if $A \supset S$ is logically true in L_0.

Relative to a list of the A-postulates of L_0, it is clear which sentences of L_0 are analytic for L_0 according to D2'. As Quine explains, the problem that concerns him lies elsewhere: "[T]he difficulty is simply that the rules contain the word 'analytic', which we do not understand! We understand what expressions the rules attribute analyticity to, but we do not understand what the rules attribute to those expressions" (1951f, 33). According to the standard reading of this passage, what a definition such as D2' does not make clear is whether the class of sentences to which the defined term "analytic for L_0" applies "is identical to the class of statements that are (in the old sense of 'analytic', whatever exactly that may be) analytic for L_0" (Becker, 83; see also Carnap 1963d, 918–919, David 1996, 1–2; and Hylton 2007, 63). This is a misunderstanding of Quine's criticism in the above passage, for two main reasons.

First, the question whether a proposed definition of "analytic for L" has the same (or a similar) extension as our already existing, historically given notion of "analytic" for sentences of L is a question of descriptive semantics, not of pure semantics. It is not relevant to Quine's investigation in TD4, where Quine is criticizing Carnap's methods of defining "analytic for L" in pure semantics, in abstraction from questions of descriptive semantics.

Second, Quine does not grant in the above passage that definitions such as D2' explain the significance of the label "analytic for L." As we saw above, an application of "true-in-L" to "Jones smokes" is as clear to us as the statement that Jones smokes. By contrast, although for any given set A of sentences of L_0, it is perfectly clear whether or not $A \supset S$ is logically true-in-L_0, this is only part of the explanation provided by D2' of what it is to be "analytic for L_0." The other, crucial, part, tacitly presupposed, is that A is the conjunction of the "A-postulates" of L_0. The A-postulates of L_0 are among what Carnap calls the "semantical rules" of L_0. And, as I noted above, by Carnap's method of defining A-true, we may select any set of sentences of L_0 to be the "A-postulates" for L_0 and, by the above method of definition, label them each "analytic for L_0." Quine's criticism is that a selection of a set of sentences as the "A-postulates" of L_0 does not explain the significance of the label "A-postulate." Such a selection therefore does not explain the significance of the label "analytic for L_0" that is defined in terms of the supposed "A-postulates" of L_0 by a rule of the form D2'. The appearance that a definition such as D2' explains the significance of the label "analytic for L_0" can therefore only be due to the occurrence of the word "analytic" in the definition.

Quine highlights this key criticism by noting that "we may, indeed, view the so-called rule as a conventional definition of a new simple symbol 'analytic-for-L_0', which might better be written untendentiously as 'K' so as not to seem to throw light on the interesting word 'analytic'" (1951f, 33). Here Quine is considering a definition such as:

D2". A sentence S of L_0 is ***K*** if and only if A ⊃ S is logically true-in-L.

Quine's criticism is that "Obviously any number of classes K, M, N, etc. of statements of L_0 can be specified for various purposes or for no purpose; what does it mean to say that K, as against M, N, etc., is the class of the 'analytic' statements of L_0?" (1951f, 33). Again, some readers (cited above) take Quine to be asking here for criteria for determining which of K, M, N, etc. is the class of analytic statements of L_0. And, again, I reject this reading on the grounds that it raises a question of descriptive semantics, not pure semantics. On my reading, in the last quoted passage Quine is asking what is distinctive about, say, the K class, as opposed to the M or N classes, each of which may also be specified by definitions that are like D2", except for presupposing different sets of "A-postulates" for L_0.[16]

Quine need not have mentioned the old word "analytic" to raise his critical question in the above passage. His mention of this word is simply a consequence of his supposition that a carnapian will take one of the sets in question, for example, the set of all and only the sentences that, by D2', are K, to be the set of "analytic" sentences of L_0. When "analytic" is taken to be just another label for the concept K defined by D2", the question, "What does it mean to say that K, as against M, N, etc., is the class of the 'analytic' statements of L_0?", becomes the question, "What is the significance of choosing the set K, as against M, N, etc.?"

To understand this challenge, it helps to recall that we can state clearly, in nonarbitrary terms, the significance of a definition of "K" that satisfies the schema (K), namely "'_____' is K if and only if _____," without presupposing or using the general notion of truth. And once we have decided to take "K" as a new definition of "true," we need not concern ourselves with questions about how to justify this decision. We may nevertheless still ask, "What does it mean to say that 'K' is the class of 'true'

---

[16] See also Quine, Letter 106 in Quine and Carnap 1990, 336–338.

sentences of L?" If we use Tarski's method of defining "K", the answer is that for each sentence of L, the definition implies a biconditional of the form (K) and each such biconditional explains in nonarbitrary, nonlinguistic terms the conditions under which K applies to that sentence. Moreover, when K is defined in Tarski's way, we can use it to generalize on sentences, as in "Every sentence of the form 'p or not p' is K." This answer does not depend on any prior meaning we associate with the word "true." In the above passage, on my reading, Quine is pointing out that we cannot similarly explain the significance of "K" as defined by D2".

This reading may seem to conflict with Quine's remark that "By saying what statements are analytic for L_0 we explain 'analytic-for-L_0' but not 'analytic', not 'analytic for'" (1951f, 33). I suggest, however, that in this sentence Quine is only granting that D2' "explains" "analytic-for-L_0" in the sense of arbitrarily fixing its extension. He is not granting, and obviously would not grant, that the "explanation" that D2' provides is of similar value or interest to the one that a Tarski's style definition of "true-in-L" provides. Hence the sentence is compatible with my reading, according to which Quine's criticism of D2" is that it does not explain the significance of the term "K" that it defines.

My reading also enables us to make good sense of Quine's next sentence, namely, "We do not begin to explain the idiom 'S is analytic for L' with variable 'S' and 'L'" (1951f, 33–34). For, as I stressed in §2, and summarized in (I') of that section, Tarski's first step toward explaining the idiom "S is true-in-L" for variable "S" and basic first-order languages "L" is to provide definitions for particular languages whose significance is as clear to us as the expressions for which they are defined. Definitions such as D2" do not explain the significance of the classes they define, so do not take an analogous first step toward solving the problem of analyticity, namely, a step of the kind I described in (I') of §1.

This criticism applies a standard of definition that Quine shares with Carnap. He and Carnap agree that to explicate an expression e that one finds useful in some respects but problematic in others (the *explicandum*) is to introduce, by definition or other methods, another expression e' (the *explicatum*) that preserves what one finds useful about e, but is clearer, more exact, or better integrated into one's current scientific theory, than e (Carnap 1947, §2; Carnap 1950b, ch. 1). For any given explicandum e, there may be more than one acceptable explicatum e'. The point of introducing an explicatum is to use it *in place of* its explicandum, so that one can retain what one finds useful about e while shedding the problems with e, by not using e at all. For this purpose, one must be able to understand and use *e* without relying on one's understanding or use of *e*.[17]

Quine's central criticism of the carnapian explicata of the first sort that Quine considers in TD4, exemplified by D2' and D2", is that they are unsatisfactory because we cannot understand and use them without relying on our understanding of the explicandum "analytic."[18] Such explicata therefore do not satisfy (I') of §3, and hence do not provide us with any clue about how to make sense of the idiom "*S* is analytic-in-*L*" for variable "*S*" and "*L*", even if we restrict the range of "*L*" to artificial basic first-order languages.

7.7 Quine's Criticisms in TD4 (Part Two)

In TD4 Quine also criticizes a second carnapian method of explaining analyticity, one that exploits the fact that "we do know enough about the intended significance of 'analytic' to know that analytic statements are supposed to be true" (1951f, 34). Quine writes,

> Let us then turn to a second form of semantical rule, which says not that such and such statements are analytic but simply that such and such statements are included among the truths. Such a rule is not subject to the criticism of containing the un-understood word 'analytic'; and we may grant for the sake of argument that there is no difficulty over the broader term 'true'. A semantical rule of this second type, a rule of truth, is not supposed to specify all the truths of the language; it merely stipulates, recursively or otherwise, a certain multitude of statements which, along with others unspecified, are to count as true. Such a rule may be conceded to be quite clear. Derivatively, afterward, analyticity can be demarcated thus: a statement is analytic if it is (not merely true but) true according to the semantical rule. (1951f, 34)

When Quine "grants for the sake of argument that there is no difficulty over the broader term 'true'," I take him to be granting, for reasons noted in §7.2, that Tarski's method shows us how to make sense of "*S* is true-in-*L*" for variable "*S*" and artificial basic first-order languages "*L*." A carnapian illustration of the second form of semantical rule that Quine

---

[17] In short, "explication is elimination" (Quine 1960, 260). Or, in other words, as Carnap explains, "The task of explication consists in transforming a given more or less inexact concept into an exact one or, rather, *in replacing the first by the second*" (Carnap 1950b, 3; my emphasis). As Howard Stein notes, "what counts in the end – still in Carnap's view of things – is the clarity and utility of the proposal [i.e. the explicatum]; whether part of that utility has to do with an earlier, vaguer, general usage is distinctly a secondary matter" (Stein 1992, 181–182).

[18] This criticism does not beg the question against Carnap and others who feel they understand the explanandum "analytic." Many informed and sophisticated readers of TD4 (e.g., Carnap 1963d, 919; Creath 2004; George 2000, 9; Hylton 2007, 63; Richardson 1998, 220) miss this crucial point.

describes, namely, one that "stipulates, recursively or otherwise, a certain multitude of statements which, along with others unspecified, are to count as true," is the following:

R. A sentence S is among the truths of L if $A \supset S$ is logically true-in-L.

Unlike the first form of rule that Quine discusses in TD4, illustrated by D2, this second form of rule does not contain the word "analytic" and therefore does not even appear to depend for its significance on a prior and independent understanding of what it is for a statement to be analytic. Also, unlike either D2' or D2'', this second sort of rule is not a definition. It simply stipulates that the sentences of L that it singles out are true. Like the Tarski-style definition of "true-in-L" it relies on, Quine thinks, a "rule of truth" such as R "may be conceded to be quite clear." Moreover, since "true-in-L" in Tarski's sense is a paradigm of a "semantical" term in the sense described at the end of §7.2, we may sensibly label R a "semantical rule."

Suppose now that we try to define analyticity-in-L in terms of a rule such as R, as follows: "S is analytic in L if S is (not merely true but) true according to the semantical rule [e.g. R]." (1951f, 34) The problem, Quine argues, is that

> Instead of appealing to an unexplained word 'analytic', we are now appealing to an unexplained phrase 'semantical rule'. Not every true statement which says that the statements of some class are true can count as a semantical rule – otherwise all truths would be "analytic" in the sense of being true according to semantical rules. Semantical rules are distinguishable, apparently, only by the fact of appearing on a page under the heading 'Semantical Rules'; and this heading is itself then meaningless. (1951f, 34)

Quine's key point is that all that we can infer from a "rule of truth," such as R, is that certain statements of an artificial language L are to count as "true-in-L according to the rule," not that the statements that are true-in-L according to the rule are different from other statements of L that are true-in-L, except for being labeled "true-in-L according to the rule." To see this, suppose we accept a certain conjunction P of the postulates of physics and we affirm

R'. A sentence S is among the truths of L if $P \supset S$ is logically true-in-L.

This rule of truth has the same form as R. Moreover, just as in the case of R, since "true-in-L" in Tarski's sense is a paradigm of a "semantical" term, we may sensibly label R' a "semantical rule." Carnap stipulates that a rule such as R picks out a special class of sentences of L, the ones that are "analytic-in-L", but a rule such as R' does not. Quine's criticism is that the act of stipulating R does not distinguish its significance from R', and hence leaves the significance of the supposed distinction between R and R' unexplained.[19]

To this criticism Carnap will of course reply that if the rule in question is a "semantical rule" in the relevant carnapian sense, then we know that the sentences that are true according to the rule are "true in virtue of meaning." In particular, Carnap will reply that since we stipulate that A is the conjunction of the A-postulates for L, we thereby determine that A, the conjunction of the A-postulates of L, and R are among the "semantical rules" of L. Hence R determines that if A ⊃ S is logically true-in-L, then S is "true in virtue of meaning." By contrast, according to this carnapian reply, since, by stipulation, P is not a conjunction of the A-postulates of L, it is not among the "semantical rules" of L, so R' does not determine that if P ⊃ S is logically true-in-L, then S is "true in virtue of meaning."

Quine's criticism of this answer is that the supposed difference between what Carnap labels a "semantical rule" and other rules that we might well call "semantical," including the supposed difference between an A-postulate and a P-postulate, and hence also the supposed difference between R and R', is not explained by Carnap's decision to label the former rules but not the other ones "semantical." It is in this context, and in this limited sense, that Quine concludes that Carnap's heading "semantical rule" is meaningless.

7.8 Two Senses of "Semantical Rule"

Quine's argument for this conclusion reveals a subtle yet important distinction between two senses of "semantical rule." In the first sense a rule is "semantical" if and only if it satisfies both of the following conditions:

(i) it specifies conditions under which a given sentence of a language has some property that we find it useful to call "semantical"; and
(ii) our application and understanding of it is independent of our inclination or decision to classify it as "semantical."

Suppose, following Tarski, we find it useful to call Tarski-style definitions of "satisfies-in-L" and "true-in-L" "semantical" on the grounds that they specify relations between objects and linguistic expressions that we may use to affirm an infinitude of sentences of L. Then Tarski-style definitions of "satisfies-in-L" and "true-in-L" satisfy condition (i). Such definitions also satisfy condition (ii), for reasons I stressed in §7.2: our

---

[19] Quine later summarized his central criticism of Carnap's analytic–synthetic distinction by saying he has found no way to distinguish between L-rules and P-rules (Quine 1963a, 398).

application and understanding of the definitions do not depend on any prior significance we associate with the defined terms, and, in particular, are independent of our inclination or decision to classify the defined terms as "semantical." Moreover, if Tarski-style definitions of "satisfies-in-L" and "true-in-L" satisfy clauses (i) and (ii), so too do R and R', each of which contains a predicate "true-in-L" defined in Tarski's way.

This unproblematic sense of "semantical rule" must be distinguished from a sense of "semantical rule" that Carnap would need to be able to explain in order to distinguish between R and R'. According to Carnap, if one decides that R, but not R', is a "semantical rule" in this supposed second sense, then one's decision implies that R partly determines the "meanings" of the sentences it applies to, whereas R' does not. Carnap's strategy of constructing definitions of L-true and A-true that satisfy his Convention AT depends on this supposed second sense of "semantical rule." My reconstruction of Quine's arguments in TD4 reveals that a "semantical rule" in this second, carnapian sense is not explained by any of the methods of pure semantics, and hence also that Carnap's special sense of "semantical" does not satisfy condition (ii).

This criticism of Carnap's use of the term "semantical rule" is an application of the same basic standards that Carnap applies when he clarifies his conception of rules and contrasts it with traditional philosophical views of rules. As I noted in §7.5, Carnap emphasizes that "The so-called rules are meant only as partial conditions of a definition" (Carnap 1963c, 923), and that the concepts of his pure semantics do not include "prescriptive components" (Carnap 1963c, 923). In 1934 Carnap proclaimed that "our own discipline, logic or the logic of science, is in the process of cutting itself loose from philosophy and becoming a properly scientific field, where all work is done according to strict scientific methods and not by means of 'higher' or 'deeper' insights" (Carnap 1934d, 46). Carnap never abandoned this commitment. It is therefore a legitimate and substantive question for Carnap whether his uses of the term "semantical rule" are clear and fully explained by his definitions.

Properly viewed, Quine's criticisms in TD4 are applications of Carnap's standards. The criticisms reveal that Carnap conflates two senses of "semantical rule," the first of which, though clear, is of no use in defining analyticity, and the second of which, though integral to Carnap's method of defining analyticity, is unexplained by Carnap's definitions. The second sense of "semantical rule" could perhaps be clarified within descriptive semantics, in mental or behavioral or cultural terms, as Quine suggests (1951f, 36).[20] Whether or not such clarifications are forthcoming, however, Quine's arguments in TD4 establish that Carnap's efforts to explicate "analytic-in-L" solely from within pure semantics in terms of what he calls "semantical rules" are unsuccessful by his own standards.[21]

---

[20] Quine takes the argument from holism of theory testing, sketched in TD5, to show that the prospects for such a clarification are dim. Carnap's and Quine's subsequent debate about the prospects for clarifying "semantical rule" in descriptive semantics (e.g., in Carnap 1955, Quine 1960, ch. 2, and Carnap 1963d, 919–921), are beyond the scope of this essay.

[21] I presented a first draft of this essay in October 2017 at a conference on Carnap and Quine organized and hosted by Sean Morris. I thank Sean for inviting me to the conference and the conference participants, including Sean, for their challenging and constructive comments. I especially thank Rick Creath, whose objections at the conference and on several later occasions convinced me that I needed to substantially revise the essay. For reading and commenting in detail on more recent drafts, I thank Kate Abramson, Joan Weiner, Sanford Shieh, Sean Morris, and Thomas Uebel.

CHAPTER 8

What Does Translation Translate? Quine, Carnap, and the Emergence of Indeterminacy

Paul A. Roth

> However, I feel no reluctance toward refusing to admit meanings, for I do not thereby deny that words and statements are meaningful .... If we are allergic to meanings as such, we can speak directly of utterances as significant or insignificant, and as synonymous or heteronymous with another. The problem of explaining these adjectives 'significant' and 'synonymous' with some degree of clarity and rigor – preferably as I see it, in terms of behavior – is as difficult as it is important. But the explanatory value of special and irreducible intermediary entities called meanings is surely illusory.[1]

What makes the meaningful meaningful – the meaning of "meaningful"? Start with some minimalist criterion of meaningfulness for a natural language – for example, "any sequence which could be uttered in the society under consideration without reactions suggesting bizarreness of idiom" (Quine 1980, 54). Given a set of sequences so generated, does this license a hypothesis that set membership must be explainable by a theory that identifies features common to what all its elements possess? That is, an underlying philosophical assumption would be that membership of any set of meaningful signs presupposes possession of some specific feature(s), and a theory of meaning explains by virtue of identifying what those are. Absent that assumption, set membership could reflect just contingent matters, accidents of time and circumstance. What made for set membership would then be indeterminate. There would be nothing for a theory of meaning to explain – no fact of the matter to be right or wrong about – because there would be no philosophical license for assuming a preexisting *common* core to the meaningful.

Carnap et al. had, of course, a very particular account on offer for purposes of theorizing meaning. Yet by virtue of rejecting as dogmas

---

[1] Quine (1948b), 11–2.

what that theory of meaning – the verifiability criterion of meaning – took as explanatia, Quine appears to then create a lacuna that needs filling, namely, to provide an alternative *theory* of meaning. This he refuses to do.[2] He offers instead a principled reason for rejecting any attempt to *theorize* meaning – the indeterminacy of translation. Statements and words he avers can be meaningful, but he denies that this fact licenses theorizing meaning. But how can one grant meaningfulness to sentences and words and yet deny legitimacy to theorizing meaning? Why Quine's steadfast conviction that explanatory appeal to "special and irreducible intermediary entities called meanings is surely illusory"?

Quine holds that accounts of analyticity and related notions lack utility for purposes of explanation. Their imagined explanatory efficacy is tied to a hope for epistemic foundations for science.[3] The twin dogmas of empiricism would make explicit inferential connections and truth conditions and so wring from ordinary language a precise delineation of the "cognitive content" of statements.[4] Meaning then becomes explanatory because it now functions as a presumptive guide to truth: analyticity (truth in virtue of meaning), synonymy (truth of the meaning equivalence of the statements or terms flanking an identity sign), and significance ("has a meaning," as truly predicated of terms or statements).[5]

Quine in "Two Dogmas of Empiricism" eviscerates any principled distinction between the meaningful and the meaningless – the sensical and

---

[2] This refusal comes out most clearly in Quine's "The Problem of Meaning in Linguistics" (Quine 1980, 47–64). Quine there rejects such theorizing as a goal. He imagines two possibilities for those interested in charting semantic information – accounting for "sameness of meaning" or "having a meaning." The first option asks after synonymy, and Quine in this regard unsurprisingly concludes "pending some definition of synonymy, we have no statement of the problem; we have nothing for the lexicographer to be right or wrong about" (1980, 63). As for "having a meaning," he suggests "treat[ing] the context 'having a meaning' in the spirit of a single word, 'significant'" (1980, 48). But about the notion of a theory of significant sequences Quine is hardly more encouraging than about synonymy. "The other [apart of attempts to specify synonymy relations] branch of the problem of meaning, namely the problem of defining significant sequence, led us into a contrary-to-fact conditional: a significant sequence is one that *could* be uttered without such and such adverse reactions. I urged that the operational context of this 'could' is incomplete, leaving scope for free supplementary determinations of a grammatical theory in light of simplicity considerations" (1980, 63–4). As discussed below, the method of using a contrary-to-fact conditional as a test for meaning turns out to be Carnap's. For a natural language, then, Quine foresees no prospect for a theory of meaning, construed as a method for *systematically* or *determinately* specifying antecedent features that bestow meaningfulness.
[3] This, of course, is the theme of Quine's (1969a), 80. I elaborate this reading in my (1999).
[4] Peter Hylton (2014b) provides a nice summary account of how this view of meaning is bred in the bone of analytic philosophy by rehearsing pre-Quinean accounts of attempts to find a principled distinction between meaningfulness and nonsense.
[5] The typology is Quine's. See his (1949).

the nonsensical – at the level of terms and statements.[6] But these very criticisms also reveal how notions of meaning embed in explanations of inferential connections.[7] This links that argument to Quine's core complaint of the nonexplanatoriness of meaning in "Truth by Convention," there in the guise of rejecting proposals that conventions can do any epistemically significant (foundational) explanatory work. "It is not clear wherein an adoption of the conventions, antecedently to their formulation, consists; such behavior is difficult to distinguish from that in which conventions are disregarded .... In dropping the attribute of deliberateness and explicitness from the notion of linguistic convention we risk depriving the latter of any explanatory force and reducing it to an idle label" (Quine 1936, 105–6).[8] Indeed, Quine's concluding observation in "Truth by Convention" – "We may wonder what one adds to the bare statement that the truths of logic and mathematics are a priori, or to the still barer behavioristic statement that they are firmly accepted, when he characterizes them as true by convention in such a sense" – makes explicit that he views as vacuous for purposes of explanation appeals to (implicit) meanings. Conversely, in order to have any explanatory import, terms or statements used as explanatia must at least invest theories using them with "deliberateness" and "explicitness." But since logic, he already notes, must be used and so presupposed in doing this, any account of logic will inevitably be fated to be assumed by a theory that uses it to explain.

But it would be a mistake to suggest that Quine was either the first or the only philosopher in a broadly analytic tradition to notice such problems. Worries about the "logocentric predicament" harken back, as Quine readily acknowledges, at least to Lewis Carroll's famous work. Moreover,

---

[6] In this regard, Quine's so-called behaviorism concerns his account of *evidence* for meaning; it does not represent any proposal for an alternative *theory* of meaning. I discuss this in more detail in my (2003) and (2019).

[7] The opening paragraph of "Truth by Convention" sounds an attack on the notion of meaning and its illusory promise of epistemically distinguishing logical symbols from other linguistic artifacts that Quine relentlessly pursues for the rest of his career. What is nascent in the critique includes worries about circularity and regress in the sense of taking logical terms as meaningful and using them in the explanatia but which in fact remain in need of explanation. Key here is appreciating the respects in which Quine in later work ties this view to a form of "essentialism." "Essentialism" and its cognate terms refer to notions that cannot be explained or explicated except by reference to related and antecedently understood notions. The clearest statement of this can be found in Section II of Quine (1963a), 387–8. By the 1960s, but, as discussed below, especially in "Natural Kinds," Quine elaborates on his reasoning for excluding such a closed circle of terms from explanation as he understands it. In Sellarsian terminology, otherwise logical terms become unexplained explainers, a type of given.

[8] Quine cites this very remark in "Carnap and Logical Truth," adding there that "It would seem that to call elementary logic true by convention is to add nothing but a metaphor to the linguistic doctrine of logical truth which, as applied to elementary logic, has itself come to seem rather an empty figure" (1963a, 392).

the philosophical status of proposed explanatia for a general criterion of meaning were already hotly debated within the core of the Vienna Circle by the 1930s.[9] In addition, and also just around the same time that Quine publishes "Truth by Convention," Charles Morris pens very similar misgivings. For in assessing how pragmatism and positivism might aid and abet one another philosophically, Morris proposes that the "concept of meaning provides the central basis upon which to discuss the present differences and tendencies" within these two movements (Morris 1936, 131). A chief complaint that Morris has with positivists as he reads them concerns how their notion of verification perpetuates "an unexamined individualistic hangover" (Morris 1936, 132) inherited from previous forms of empiricism.

Against this, Morris urges that "Verification is in general a matter of more or less, and in scientific usage the term knowledge includes all those meanings which through the social process of verification have come to act as relatively stable points of reference in scientific procedures" (Morris 1936, 133). Given this more expansive and social view of verification, he further concludes that "it is clear that there is no absolute line of demarcation to determine when a proposition is to be given an honorific status in the domain of knowledge" and that "there can be no certainty that the rank of any specific propositions will henceforth undergo no change" (Morris 1936, 134).[10] Morris makes this Quinean-sounding point even more explicitly a year later:

> Indeed, in the very admission that conventional factors and logical structure are to be discriminated from empirical investigations, is there not an admission that not only pragmatism and formalism are transcended, but empiricism as well? ... Thus the contrast of the logical and the empirical, real enough at any moment, is a relative distinction .... Logic as the science of logical analysis can without inconsistency be an empirical science. (Morris 1937/1979, 52)

Concerns about how appeals to meaning "taint" claims to explicate or to explain, including any presumed divide between the empirical and the logical, cannot be said to be Quine's alone.

---

[9] Thomas Uebel's work indicates that this debate rages during the late 1920s and early 1930s. See, e.g., his (2004). For a fuller elaboration, see his (2007a).

[10] In this spirit, Morris also observes that "when we choose the rules of operation for a constructed language, we must understand in a non-formal sense what operations the rules permit. In both cases we can later formulate the rules themselves in formal terms, but only by using language not itself at that moment in the purely formal mode" (Morris 1936, 137).

But just as contemporaries shared Quine's doubts about any possibility of a principled separation of the logical and the empirical for purposes of explanation, worries that problems regarding a verifiability criterion for individual terms and statements collapse into holism also prove to be harbored by others in the near philosophical vicinity. In an article by Hempel that Quine cites in "Two Dogmas" (1980, 37, fn. 15),[11] Hempel acknowledges the following consequence of his (Hempel's) review of various failed attempts at a criterion of meaning:

> Hence, what is sweepingly referred to as "the (cognitive) meaning" of a given scientific hypothesis cannot be adequately characterized in terms of potential observational evidence alone, nor can it be specified for the hypothesis taken in isolation: In order to understand "the meaning" of a hypothesis within an empiricist language, we have to know not merely what observation sentences it entails alone or in conjunction with subsidiary hypotheses, but also what other, non-observational, empirical sentences are entailed by it, what sentences in the given language would confirm or disconfirm it, and for what other hypotheses the given one would be confirmatory or disconfirmatory. *In other words, the cognitive meaning of a statement in an empiricist language is reflected in the totality of its logical relationships to all other statements in that language and not to the observation sentences alone.* In this sense, the statements of empirical science have a surplus meaning over and above what can be expressed in terms of relevant observation sentences. (Hempel 1950, 59; emphasis added)

Interestingly, Quine references Hempel only regarding related criticisms of the verification criterion of meaning. It is not until later in "Two Dogmas" that Quine writes, "My countersuggestion ... is that our statements about the external world face the tribunal of sense experience not individually but only as a corporate body [citing Duhem]" (1980, 41). He goes on to punctuate this point with his much-cited remark that "The unit of empirical significance is the whole of science" (1980, 42). But as Quine obviously knows, Hempel has already penned words apparently to the same conclusion but which Quine at just this juncture seems to ignore.

Situating Quine relative to Morris and Hempel in this way serves to underscore a certain exegetical question: What did Quine take himself to be contributing to the debate about meaning? My question does *not* reflect some doubt about Quine's own importance. Rather, it invites a reconsideration of what to take as Quine's target in tracking the evolution

---

[11] This footnote, as well as the reference to Duhem mentioned below, were added when the essay was reprinted in the 1961 edition of (1980). However, Quine nowhere indicates that he takes Hempel to have anticipated his point.

of his debate with Carnap that crystallizes in the indeterminacy thesis. The historical context suggests that this dispute must at its core turn on something other than critiques of empiricist dogmas already broached by others. In this fundamental regard, uncovering what troubles Quine about proposed theories of meaning must go beyond points related to issues already emphasized in the philosophical literature and quite familiar to disputants on all sides.[12]

Quine, I claim, ultimately realizes that the only explanatory function a theory of meaning could serve would be to assume the presence of *antecedent* features that then permit a systematic determination delimiting the set of meaningful terms and meaningful statements. Such a theory explains because it provides nonaccidental, indeed determinate reasons for membership of that set. This specific assumption therefore attributes a special metaphysical status to meaning. It is metaphysical because it takes such features as in place prior to any empirical inquiry and in a way that can be tested for but not disconfirmed by evidence.

Thought of in this way, two points made by Peter Hylton in his otherwise estimable book on Quine must then be incorrect. Hylton claims that indeterminacy is neither a central doctrine of nor crucial for Quine's overall position. Hylton writes that the amount of discussion devoted to indeterminacy by Quine, as well as by Quine's critics, might suggest that "indeterminacy is a central Quinean doctrine" (Hylton 2007, 200). These implications Hylton denies. He asserts instead that "the idea is of less importance to Quine than is often supposed, and is not crucial for his thought in general" (Hylton 2007, 200; see also 226).[13] But Hylton's claims

---

[12] For a detailed and thoughtful development related to this line of interpretation, see Paul Livingston, "Quine's Appeal to Use and the Genealogy of Indeterminacy," ch. 5 of his (2008).

[13] In this respect, Hylton relegates the indeterminacy thesis to the status of a mere "conjecture" offered by Quine, claiming this commensurate with Quine's own later, more considered view. Textual support for this attribution is tenuous. For example, in Quine's brief 1994 piece, "Indeterminacy Without Tears," Quine does write of his having "hazarded the following indeterminacy thesis," which he then goes on to put in familiar terms of competing manuals of translation (1994c, 447). He refers to this thought experiment as his "conjecture." But in context, what is conjectural is the hypothesized existence of competing manuals, a conjecture meant only to illustrate the thesis. "The philosophical point of my conjecture is the untenability of the notion of propositions, or sentence meanings" (1994c, 447). This becomes clear when Quine writes immediately after speaking of his "conjecture" that "In stating the thesis just now I invoked a relation of equivalence, applied to sentences within our own language. I must eliminate this, for it is too close to the notion of sentence meanings that I am challenging" (1994c, 447). The "conjecture" hypothesizes, in other words, equivalent manuals based "on the limitations of possible evidence in the native's verbal behavior and its observable circumstances" (1994c, 447). So the conjecture refers to the thought experiment; it illustrates the thesis but is not the argument for it. Indeed, and as discussed in the body of my essay, Quine consistently writes of the "thesis" of the indeterminacy of translation. Writing just a few years before the piece that Hylton cited, Quine unhesitantly uses the unequivocal characterizations

that the indeterminacy thesis is neither central to nor crucial for Quine's overall position cannot be squared with Quine's own writing about and positioning of indeterminacy. The second and longest chapter of *Word and Object* Quine devotes to laying out a case for the indeterminacy thesis, writing there of "a principle of indeterminacy of translation" (Quine 1960, 27). Likewise, writing about a decade later in "Epistemology Naturalized," Quine situates discussion of indeterminacy immediately subsequent to that of his critiques of Carnap and foundationalism. Quine there states explicitly that these "considerations raise a philosophical question even about ordinary unphilosophical translation" (1969a, 80). In short, whatever points Quine takes to count against Carnapian translation count as well against "ordinary unphilosophical translation."

Just a little over a page later, Quine writes that "The crucial consideration behind my argument for the indeterminacy of translation was that a statement about the world does not always or usually have a separable fund of empirical consequences to call its own" (1969a, 82). This echoes the point Quine makes against Carnap's notion of definition that Quine discusses just prior to the passage quoted. And in terms that recall Hempel's remark about "surplus of meaning," Quine declares that "In giving up hope of such translation, then, the empiricist is conceding that the empirical meanings of typical statements about the external world are inaccessible and ineffable" (1969a, 78–9). Contrary to what Hylton says then, arguments supporting indeterminacy are at one with those central to Quine's critique of Carnap in particular and semantic theory in general. Indeterminacy is a direct consequence of what Quine takes to be his core critique of the notion of meaning.

As has already been noted, the brunt of Quine's misgivings focus from the first on the absence of any possible *explanatory* efficacy of a theory of meaning by way of systematically accounting for a set of linguistic units identified as meaningful. To develop the worry, however, requires teasing out what a theory of meaning presupposes that other forms of inquiry that Quine will accept as scientific do not.

---

that one finds starting with the late 1950s work in which indeterminacy first appears. "The critique of meaning level by my thesis of indeterminacy of translation is meant to clear away misconceptions, but the result is not nihilism. Translation remains, and is indispensable. Indeterminacy means not that there is no acceptable translation, but that there are many" (1987, 345). Indeterminacy accounts for why there must in principle be many; the thought experiment of competing manuals illustrates why. The principled possibility exists because "there is simply no fact of the matter .... Translation *remains* indeterminate, even relative to the chosen theory of nature. Thus the indeterminacy is an indeterminacy *additional to* the underdetermination of nature" (1987, 346; emphasis added). I can find no evidence that Quine retreats from this principled characterization.

This emphasis on the explanatory project as Quine conceives of it has some affinities with conclusions reached in Gary Ebbs' searching examinations of the Quine–Carnap debate.[14] Like Ebbs, I find the basic animus of Quine's critique of Carnap to lie with the explanatory role that Quine takes notions such as analyticity to have in Carnap's scheme. (See, e.g., Ebbs 2017, 59.) But Ebbs diagnoses the explanatory shortfall Quine finds in Carnap as symptomatic of Quine's nascent naturalism. On this reading, Quine objects to Carnap's account of analyticity because it imports an extra-scientific form of justification. "The problem is that Carnap's definition of 'analytic-in-LS' is designed to explicate a conception of justification for accepting statements that is independent of the statement's explanatory contribution to scientific theory – a conception of justification that Quine associates with first philosophy" (Ebbs 2017, 81, fn. 27). But this characterization presumes that Quine provides some precise account of when a statement could or could not make an "explanatory contribution to scientific theory." Quine does not. This forces the question, then, of why Carnap's position could not be accommodated within Quine's.[15]

Indeed, at other points, Ebbs appears to acknowledge this. For as Ebbs notes, truths that "may be called 'explanatory' in one good, though very broad, sense of that word … are just the ones that it is the aim of science to affirm. It follows that we have no grip on what is to be 'explanatory' in this encompassing sense apart from our own ongoing scientific inquiries" (2017, 81). But now it appears that Ebbs' Quine has just begged the question against Carnap's account. For the question as Ebbs recognizes is supposed to be how any proposed postulates fit with the larger project of inquiry. All agree in the end that logic does fit. So Ebbs' account renders true but trivial Quine's complaint that Carnap's view is "explanatorily empty" (Ebbs 2017, 81, fn. 17). The triviality resides in the fact that a minor modification now appears to be all that separates Quine and Carnap, one that should have been apparent to both.[16]

---

[14] Ebbs' several essays on this topic have been recently collected in his (2017).

[15] Here and throughout this essay I concern myself only with how I understand Quine to have read Carnap. Whether Quine's reading is the best reading of that material I leave unaddressed here. As quotations provided towards the end of this essay demonstrate, enough contemporary philosophers subscribe to the view that Quine imputes to Carnap to make that view of independent philosophical interest. The indeterminacy thesis has then substantive philosophical import apart from any exegetical question of how best to read Carnap. I thank Thomas Uebel for pushing me to clarify this.

[16] A sense in which Quine must have felt like he was shouting into the wind can be heard in his following remark: "Carnap maintains that ontological questions, and likewise questions of logical or mathematical principle, are questions not of fact but of choosing a convenient conceptual scheme or framework for science; and with this I agree only if the same be conceded for every scientific hypothesis" (Quine 1951b, 72). An important exegetical puzzle my reading solves thus consists in identifying why Quine takes Carnap to resist this characterization.

No small part of the problem then, and one left unaddressed by Ebbs, is that Quine simply offers no principled distinction between science and nonscience. Invoking the "aim of science" consequently can be of no help, especially since Quine has a rather liberal conception what to count as a science.[17] Thus if pressed about why *in principle* truths of meaning and the like *cannot* be part of science, Ebbs has no answer on Quine's behalf.

This comes out even more clearly by tracing an important question that Ebbs raises in his reading of Quine: "Are some parts of our scientific theory both explanatory and true by convention?" (2017, 87). Ebbs claims that the "standard story" of the Quine–Carnap debate has it that Quine's answer to this question is no (2017, 87). Ebbs then goes on to argue, correctly I would say, that for Quine "there is at most a difference in degree to which conventional, pragmatic choice plays a role in our adoption of a legislative postulate, on the one hand, and our adoption of an empirical hypothesis on the other" (2017, 90). In words that recall what Morris wrote, Ebbs then asserts that "The set of sentence-affirming events that are conventional for us at one time may therefore be different from the set of sentence-affirming events that are conventional for us at another time" (2017, 90). From this, Ebbs plausibly concludes that Quine requires only that postulation "be regarded, at the time of the act, by the person whose act it is, as explanation in the broad sense I sketched above" (2017, 91). Yet this simply circles back to Ebbs' own characterization of Quine's view, to the effect that "we have no grip on what is to be 'explanatory' in this encompassing sense apart from our own ongoing scientific inquires" (2017, 81). But again, inasmuch as Quine offers no demarcation criterion for a science, any attempt to justify a *principled* exclusion of theories of meaning by appeal to "scientific inquiries" must be a nonstarter. As a consequence, nothing Ebbs offers by way interpretation and defense of Quine illuminates why the notion of meaning must be regarded from the outset as explanatorily empty.

Ebbs' claim that he presents "Quine's criticisms of his version of the thesis that logic is true by convention … in a way that sheds light on

---

[17] I discuss Quine's fluid conception of what can be a science in "The Epistemology of 'Epistemology Naturalized'," esp. 106f. See also my (2006) and (2008). For reasons beyond the scope of this essay to argue, Ebbs also wrongly takes Quine's naturalism to be critical to the argument for the indeterminacy. But apart from the fact that Quine nowhere cites naturalism as a premise, he positions his naturalism as a consequence of the arguments that suffice for indeterminacy. I have elsewhere argued for the independence in Quine of his holism and his naturalism. See my (1984). Reading Quine problematizing the notion of a linguistic framework and its use in underwriting a certain conception of scientific rationality echoes important themes identified and developed by Thomas Ricketts in his (1982). However, my account of indeterminacy diverges from that which Ricketts offers there.

his use in those criticism of the words 'explanation' and 'explanatory'" (2017, 91–2). But given Quine's own acknowledgment that his criticism is rooted in the sort of regress problem famously noted by Lewis Carroll (see Quine 1963a, 391, fn. 4), Ebbs' interpretative efforts here seem at best redundant. Conventions are not self-interpreting; in that sense explanation by appeal to conventions in logic leaves explanation unbegun. But since logic *can* be part of science, regress problems notwithstanding, this circularity cannot then justify exclusions of postulates related to theories of meaning. *Abetting explanation encourages postulation; integrating postulates into a broadly naturalist framework certifies them as genuinely explanatory.* The notion of meaning, Quine insists, cannot satisfy the latter constraint, whatever else it promises. The question that Ebbs never answers on Quine's behalf then is why meaning cannot be made naturalistically respectable.

Ebbs and I do agree, however, that when it comes to accounting for indeterminacy, the issue turns on Quine's claim that a theory of meaning cannot be explanatory. "Quine's account of meaning and his thesis of the indeterminacy of translation do not amount to a substantive theory of sentence meaning, but to a denial of the appropriateness of theorizing about sentence meaning" (Ebbs 2017, 118). But having written this, Ebbs then attributes to Quine the view that "we should focus instead on formulating clear, useful, and explanatory theories" (2017, 118), including, it turns out, Ebbs' reading into Quine a Quinean theory of meaning. "What is less well known is that Quine explicitly endorses a method for this purpose" (Ebbs 2017, 118), namely, "the 'method of explication'" (2017, 118). Indeed, Ebbs goes so far as to claim that Quine's "indeterminacy thesis depends on his explication of meaning as what is determinate by the totality of speech dispositions" (2017, 119). But this reading of the argument for indeterminacy has no textual support. None of Quine's own sketches of the argument cite explication as a premise. And good reason exists for resisting on Quine's behalf this reading of the argument for the indeterminacy, since it simply begs the question against a theory of meaning by assuming it can be no more than "what is determinate by the totality of speech dispositions." Quine premises his claim that "Surely one has no choice but to be an empiricist so far as one's theory of linguistic meaning is concerned" (1969a, 81) on the indeterminacy argument; it cannot be a premise for it.

The indeterminacy thesis is thus not a secondary feature of Quine's position (as Hylton maintains). Nor does Quine's notion of a science commit him to a limit on posits, to an especially restrictive account of evidence, or to a conception of explanation that clearly excludes appeals to meaning (as Ebbs infers). The failure of a theory of meaning to be explanatory cannot be tied to the fact that such a theory lies outside science, since the case for what is or is not a science presupposes an account of explanatory efficacy. "When on the other hand we take the verification theory of meaning seriously, the indeterminacy of translation would appear to be inescapable. *The Vienna Circle espoused a verification theory of meaning but did not take it seriously enough*" (1969a, 80; emphasis added; see also 1960, 79). The question thus remains: What precludes a theory of meaning from being a part of science? Why does Quine expect one to agree, as surely people do, that terms and statements can be meaningful and yet deny that a theory of meaning has any role to play for purposes of explanation?

A key to understanding what is original in Quine's criticism involves unpacking Quine's remark from "The Problem of Meaning in Linguistics" that "in the case of the lexicon" (i.e., a catalog of meanings) there exists "nothing for the lexicographer to be right or wrong about" (1980, 63). This echoes later formulations that in the case of indeterminacy, "there is no fact of the matter" (Quine 1969f, 303). Why is it that what suffices as the "unit of empirical significance" for science fails nonetheless to provide a fact of the matter for translation?

In the period between 1951 and 1960 – between the publication of "Two Dogmas" and that of *Word and Object* – Quine comes to realize that his rejection of the analytic–synthetic distinction implies a rejection of the very possibility of a theory of meaning. One characteristically quotable Quinean line from "Two Dogmas" hints at this: "Meaning is what essence becomes when it is divorced from the object of reference and wedded to the word" (Quine 1980, 22). Why "essence"? Quine goes on to argue that "the dogma of reductionism has, in a subtler but more tenuous form ... [been] of course implicit in the verification theory of meaning" (1980, 40–1). He then argues that *both* the analytic–synthetic distinction and reductionism qua dogmas "are, indeed, at root identical" (1980, 41). The essentialism resides in the fact that imputed necessary equivalences – either verbal or empirical – appear presupposed in a way that cannot be *empirically* justified.

This nonempirical presupposition makes such relations metaphysical. Thus when Quine envisions an "Empiricism Without the Dogmas" in Section VI of that essay, his alternative imagines an empiricism *absent* any theory of meaning that would guide inquiry by referencing necessary connections: "I envisage nothing more than a loose association reflecting the relative likelihood, in practice, of our choosing one statement rather than another for revision in the event of recalcitrant experience" (1980, 43). *Empiricism without dogmas is an empiricism without a theory of meaning.*

Unlike other forms of inquiry that Quine allows within his transformed account of science, no analog to a theory of meaning reappears within the web of belief.

Failure to consider Quine's complete rejection of a theory of meaning might appear to make plausible a suggestion that Quine's critique of empiricist metaphysics culminating in "Two Dogmas," aimed as it seems to be at a general theory of meaning, leaves untouched the more moderate strategy that Carnap offers in "Empiricism, Semantics, and Ontology."[18] Yet Quine continues to resist. And rightly so, for to so read Quine would be to miss the force of his criticism, one that aims at any form of essentialism against which undogmatic explanation must guard. Empiricism as Quine begins to reimagine it in the last section of "Two Dogmas" hints that this precludes theories of meaning. What he comes to appreciate, or so I argue below, is that more needs to be said.

The same year he publishes "Two Dogmas" he also publishes "On Carnap's Views on Ontology."[19] Here Quine begins to make more explicit concerns he harbors regarding connections between analyticity and ontology. These issues he writes there "prove to be interrelated" (1951b, 65). As he notes, "Carnap thinks – and here is a more than terminological issue – that the question what a theory presupposes that there is should be divided into two questions in a certain way; and I disagree" (1951b, 66). This all sounds familiar, except Quine adds a certain philosophical fillip: "Now to determine what entities a given theory presupposes is one thing, and to determine what entities a theory should be allowed to presuppose, what entities there really are, is another. It is especially in the latter connection

---

[18] Huw Price reads Quine this way. Price sees Quine's "good" criticism of Carnap turning on a failure by Carnap to provide a technical solution to distinctions between frameworks. "But he [Carnap] does little to defend the assumption that the boundaries are there to be marked – and this is what Quine denies. Tradition seems to assume that Quine has an argument for the opposing view – an argument for *monism*, where Carnap requires *pluralism*. I want to show that this is a mistake, and rests on a confusion of two theoretical issues concerning language" (Price 2009, 289). The supposed "theoretical confusion" imputes to Quine a failure to appreciate *functional* differences in the ways that different parts of language get used. "In virtue of the preexisting functional differences between these concepts with which they associate, however, the different applications of these terms are incommensurable ... Nothing in Quine's criticisms of Carnap's and Ryle's pluralism seem to count against the existence of such foundations, and so the verdict on the Carnap-Ryle view must await ... first-order scientific inquiries into the underlying functions of language in human life" (Price 2009, 252–4). This same basic argument appears in Price's (2007), 396ff. But Price just has it wrong, because he completely misses what Quine objects to when objecting to a theory of meaning. As shown below in the discussion of the "recapture" view, the "functional pluralism" Price invokes entails precisely the assumption Quine takes as untenable, i.e., that of a "preexisting functional difference" as even a possibility to be uncovered by scientific inquiry.

[19] Since he references the published version of "Two Dogmas" in "On Carnap's Views on Ontology," I assume that the former is written before the latter.

that Carnap urges the dichotomy which I said I would talk about" (1951b, 67). But what does Quine imply when he writes of what entities a theory "should be allowed to presuppose"? The formulation sounds unQuinean. What prohibits a theorizer from postulating for purposes of explanation?

Quine illustrates his point by giving a long quotation from Carnap's original publication of "Empiricism, Semantics, and Ontology" wherein Carnap delineates his distinction between internal questions and external questions. Quine edits the quotation to focus on Carnap's use of the distinction for purposes of arbitrating ontological disputes. In questioning whether or not Carnap can distinguish in a principled way between what makes for an internal question, Quine considers and rejects various technical strategies by which theoretical standpoints might be distinguished. Then near the very end of his paper he announces, "the basic point of contention has just emerged: the distinction between the analytic and the synthetic itself" (1951b, 71). Quine then passingly refers to "Two Dogmas," remarking that he "will not retrace those steps here" (1951b, 71). The added fillip, then, is this: "If there is no proper distinction between analytic and synthetic, then no basis at all remains for the contrast which Carnap urges between ontological statements and empirical statements of existence" (1951b, 71). How is this additional to what is already argued in "Two Dogmas"? Immediately prior to this remark, Quine quotes Carnap from "Empiricism, Semantics, and Ontology":

> Quine does not acknowledge the distinction which I emphasize above [viz. the distinction between ontological question and factual questions of existence], because according to his general conception there are no sharp boundary lines between logical and factual truth, *between questions of meaning and questions of fact*, between *acceptance of a language structure* and the acceptance of an assertion formulated in the language. (1951b, 71; emphasis added)

Unlike "Two Dogmas," Quine now realizes that his position entails precluding the very possibility of formulating *any* separate framework for meaning of the sort Carnap imagines. *Carnap's distinction presupposes that a notion of a language structure makes sense prior to beginning inquiry; it is this assumption that Quine rejects.* He rejects it not just because of concerns about analyticity, but because Carnap's assumption imports without argument a notion of prior structure, one philosophically robust enough to license a distinction between, *inter alia*, a structure in use and one hypothesized for purposes of inquiry.

As already noted above, Quine's envisions an empiricism *sans* theory of meaning. What remains nascent concerns why Quine's position precludes theories of this sort. The italicized remarks highlight that Quine's now emphasizes more what an account of "language structure" presumes. What begins to emerge here concerns, or so I argue, an appreciation that a theory of meaning presupposes a more general assumption that all meaningful elements stand in a systematic relation *before* translation begins. If they did not, no theory could determinately explain meaningfulness *per se*. The previously identified dogmas of empiricism can do explanatory work only if systematically linked in a way that translation might recapture. Absent that assumption, a theory of meaning has no work to do.

The putative explanatia of such a theory explained meaningful sequences – equivalence and significance in particular – by virtue of making explicit what made them meaningful in specific respects. Since the problems of holism, circularity, and regress will, for Quine, turn out to be inescapable features of *any* theory, his reason for ruling out theories of meaning from science cannot straightforwardly be identified with these features.[20]

In "The Problem of Meaning in Linguistics," where clear precursors to the indeterminacy thesis occur, Quine emphasizes that in order for a theory of meaning to explain meaningfulness, the meaningful must be assumed to be a product of some *prior* systematic relationship. He imagines there in anticipation of his field linguist a grammarian struggling with "a hitherto unstudied language" (1980, 49) and attempting to develop "a class K of significant sequences of the language" (1980, 49) for purposes of finding their synonymous equivalents in English. But for there to be this class, the linguist must assume "that the class K is *objectively determinate before the grammatical research is begun*" (1980, 51, emphasis added). Formal implication makes for determinacy, and explicit rules of implication make

---

[20] Here it helps to examine a strategy Quine employs in his essay "Natural Kinds." His remarks there on the notion of similarity and its role for purposes of scientific explanation help clarify why a theory of meaning can have no explanatory purchase. "Disposition terms and subjunctive conditionals in this area, where suitable senses of similarity and kind are forthcoming, suddenly turn respectable; respectable and, in principle, superfluous. They may be seen perhaps as unredeemed notes; the theory that would clear up the unanalyzed underlying similarity notion in such cases is still to come .... In general we can take it as a very special mark of the maturity of a branch of science that it no longer needs an irreducible notion of similarity and kind .... In this career of the similarity notion, ... we have a paradigm of the evolution of unreason into science" (1969b, 138).

Quine explicitly implicates Carnap in failed efforts to explicate or define similarity in logical terms (1969b, 120–1). The reference is hardly gratuitous, but points back to a fundamental aspect of Carnap's continued efforts to theorize meaning. Terms like "natural kind" or "similar" when invoked supposedly in the service of explanation connote distinctions that obtain *prior* to any science. Because of this, a science that uses such notions does not also explain them. *If not explained by science, then not fully integrated into a reasoned understanding of the world. If explained, then superfluous.*

for objectivity.[21] Why must the investigator assume *prior* to research this objective determinacy? Absent that assumption, there would be no *objectively determinate* class K for which to find synonyms.

Quine's conjectured investigator "wants to devise, in terms of elaborate conditions of phoneme succession alone, a necessary and sufficient condition for membership in K. He is an empirical scientist, and his result will be right or wrong according as he produces that objectively predetermined class K or some other" (1980, 52). This assumption of the prior existence of a determinate class K Quine explicitly takes to be necessary for assuming that the grammarian qua inquirer has a subject matter – literally, a fact of the matter – to investigate.

> But the setting of the grammarian's problem is quite another matter, for it turns on a *prior* notion of significant sequence, or possible normal utterance. Without this notion or something somewhat to the same effect, we cannot say what the grammarian is trying to do – what he is trying to match in his formal reproduction of K – nor wherein the rightness or wrongness of his results might consist. (1980, 52, emphasis added; see also 54)

And it is in the acceptability of this precise assumption – setting the problem as one of recapturing a prior (to theorizing) notion of significant sequence – that he rejects. Rejection leads Quine there to conclude, in anticipation of his "no fact of the matter" formulations of indeterminacy, that "In the case of the lexicon, pending some definition of synonymy we have no statement of the problem; we have nothing for the lexicographer to be right or wrong about" (1980, 63). In order to have a theory that explains or accounts for synonymy relations by a theory of meaning, one must assume that statements or terms constituting the explananda stand in some prior systematic relation both to one another and to potential explanatia. If there is no prior systematic relation, and so no way to objectively generate the class K of significant strings, then there can be no formulation of a problem to be solved, nothing for a theory of meaning to be a theory of.

This does not beg the question against meanings. Rather, Quine asks for some way of establishing that all the elements of K are members of the set only because they share a common something that bestows them with meaning. Meaningfulness is not enough; for that a mere behavioral standard suffices. The problematic move involves inferring that granting meaningfulness entails some determinate feature(s) to be explained by a

---

[21] Building on work by Thomas Ricketts, I develop in Roth (2019) an account of how the term "indeterminacy" is best understood in relation to what makes for determinacy as discussed in Carnap. See also Livingston's important discussion of this topic in Livingston (2008), ch. 5.

theory of meaning. The metaphysical assumption that Quine thus resists presumes that meaningfulness must be a consequence of such features. In rejecting the dogmas of empiricism, Quine has already removed presumptions of a common core. To insist that there yet must be one is not to rule meaning out, but to ask for a basis for ruling it back in. Granting that there are meaningful utterances is as yet no reason for postulating any "common core."

Carnap, of course, does not take Quine's point, and Carnap's theorizing about intensional content only deepens the divide. For Carnap continues to theorize about meaning, and to do so with a vengeance. He appears fully cognizant that Quine looks to preclude theories of meaning *tout court*. Whatever Carnap might not understand about the force of Quine's criticism, he does appreciate that a key point separating his philosophical approach and Quine's involves just this assumption of prior systematicity.

Carnap has Quine squarely in his sights in this specific regard in "Meaning and Synonymy in Natural Languages." Carnap's choice of target is of particular note, for the piece by Quine on which he focuses is not "Two Dogmas" but "The Problem of Meaning in Linguistics." Carnap obviously notes Quine's passing use of an image of a grammarian "at work on a hitherto unstudied language" whose "own contact with the language has been limited to his field work" (1980, 49). Carnap appropriates this translation trope and uses it to his own ends.[22]

Carnap read "On the Problem of Meaning in Linguistics," "On Carnap's Views on Ontology," and "Carnap and Logical Truth" prior to writing "Meaning and Synonymy in Natural Languages," and it is precisely the assumption of prior systematicity that he explicitly reasserts against Quine. "The theory of the relation between a language – either a natural language or a language system – and what a language is about may be divided into two parts which I call the theory of extension and the theory of intension" (1955, 33). Carnap identifies intension and meaning, but what is most important for purposes of understanding Quine's critique is that Carnap takes natural as well as formal languages to embody systematic

---

[22] Quine, of course, also uses the translation trope in "Carnap and Logical Truth" when talking about an "imaginary positivist" he there dubs Ixmann. Quine remarks in a footnote to the version of (1963a) published in 1960 ("Carnap and Logical Truth," *Synthese* (1960) 12:350–74, 350, fn. 1) that he completed the essay in 1954 for what turned out to be the long-delayed volume on Carnap edited by Schilpp, which did not come out until 1963. But Carnap had in fact read the essay by 1954, as an exchange of letters between Carnap and Quine from July and August of 1954 documents. See Quine and Carnap (1990), 435–9. I speculate that Carnap's appropriation of this trope in "Meaning and Synonymy" goads Quine to further develop his thoughts about a field linguist, radical translation, and so indeterminacy over the next few years.

interrelations, namely, a way of specifying prior to inquiry Quine's class K. Indeed, Carnap endorses this assumption in a strong form:

> From a systematic point of view, the description of a language may well begin with the theory of intension and then build the theory of extension on its basis. By learning the theory of intension of a language, say German, we learn the intensions of the words and phrases and finally of the sentences. Thus the theory of intensions of a given language L enables us to understand the sentences of L. (1955, 34)

Carnap's later position, of course, represents a shift away from the sort of theory of meaning he earlier advocates. Nonetheless, he retains unaltered his view that the meaningful segments of a natural language constitute explananda for a theory of meaning, a theory that accounts for how a language generates that very class of meaningful terms and statements.

Carnap, well aware of Quine's criticisms (he references both "Two Dogmas" and especially "The Problem of Meaning in Linguistics" in his essay at (1955), 36, fn. 3 and the latter again at 37), takes up his understanding of Quine's challenge and asks how a linguist could "go beyond [determinations of extension] and determine also its intension?" (1955, 36). In this regard, Carnap does not hesitate to speak of meaning as the "cognitive component" of a statement, one which informs on its truth. That is, he makes determination of truth conditions a *consequence* of the intensional component. "The technical term 'intension,' which I use here instead of the ambiguous word 'meaning,' is meant to apply only to the cognitive or designative meaning component. I shall not try to define this component. It was mentioned earlier that determination of truth presupposes knowledge of meaning" (1955, 37). Just here Carnap conjures up his own case of translation, with competing linguists speculating about the language of Karl.

Carnap readily acknowledges that determination of extension might well leave intensional content undetermined (or underdetermined):

> As long as only these results [determinations of extension] are given, no matter how large the region is – you may take it, fictitiously, as the whole world if you like – it is still possible for the linguists to ascribe to the predicate different intensions. For there are more than one and possibly infinitely many properties whose extension within the given region is just the extension determined for the predicate.[23] (1955, 37)

Carnap correctly notes that "Here we come to the core of the controversy. It concerns the nature of a linguist's assignment of one of these properties

---

[23] Quine seems to deliberately echo this wording in his (1970c), 179, 180.

to the predicate as its intension" (1955, 37). So, having already conceded that nothing actually empirically available can help at this juncture, what does he imagine can determine this assignment?

To develop the class K of significant sequences, the investigator "must take into account not only the actual cases, but also possible cases .... All logically possible cases come into consideration for the determination of intensions" (1955, 38). "Two Dogmas" notwithstanding, Carnap still works with the assumption that meaningfulness implies the presence of a theory of meaning:

> The man on the street is very well able to understand and to answer the questions about assumed situations ... I believe that it is sufficient to make clear that it would be possible to write along the lines indicated a manual for determining intensions or, more exactly, for testing hypotheses concerning intensions. The kinds of rules in such a manual would not be essentially different from those customarily given for procedures in psychology, linguistics, and anthropology. (1955, 40)

Quine will later come to disparage what he terms a system of analytical hypotheses,[24] as well as his subsequent dismissal of notions of kind and similarity noted above. But here Carnap soldiers on with his unargued for assumption of systematicity. "That X is able to use a language L means that X has a certain system of interconnected dispositions for certain linguistic responses" (1955, 42). Now Carnap imagines that this set of dispositions might be teased out behavioristically or, in some future state of more detailed physiological knowledge, by means of what Carnap terms a method of "structure analysis." Here significant sequences supervene on physical states in a way that provide a type–type identity.

> On the basis of the given blueprint of X, he [an investigator] may be able to calculate the responses which X would make to various possible inputs. In particular, he may be able to derive from the given blueprint, with the help of those laws of physics which determine the function of the organs of X, the following result with respect to a given predicate 'Q' of the language L of X and specified properties F_1 and F_2, (observable for X): If the predicate 'Q' is presented at C, then X gives an affirmative response if and only if an object having the property F1 is presented at A and a negative response if and only if an object with F2 is presented at A. This result indicates that the boundary of the intension of 'Q' is somewhere between the boundary of F1 and that of F2. For some predicates the zone of indeterminateness between F, and F2 may be fairly small and hence this preliminary determination of the intension fairly precise. (MS 44)

[24] See Quine (1959), esp. section 6; cp. Quine (1960), ch. 2, §15.

Carnap, while acknowledging a possible "zone of indeterminateness" in ascertaining the boundary of application, nonetheless persists in maintaining that this "method of structure analysis" "can supply a general answer and, under favorable circumstances, even a complete answer to the question of the intension of a given predicate" (1955, 44). Carnap, moreover, does not at all take it that the hypothetical person under study provides information only about that individual's meanings. Otherwise, like the situation that Davidson (1986) imagines in "A Nice Derangement of Epitaphs," there would be no licensing an assumption of a theory of meaning shared by all speakers of a language. Rather, Carnap takes the system of dispositions of an interrogated speaker to be representative. In this key respect, Carnap's Karl goes proxy for all speakers of that language, and so too for the systematic dispositions displayed.

On my reading, what Carnap calls a "hypothesis of intension" can only have a fact of the matter, an objective something to theorize about, if there exists a boundary to discover, one somehow in place prior to inquiry. The delineation of a class K that includes counterfactuals, as Carnap proposes, is possible only on such an assumption. I take this to be why Carnap acknowledges that this is "the core of the controversy." His 1955 discussion strongly hints at how such an assumption operates in his account. For immediately after considering the linguists who puzzle over how to determine the intensions of Karl's concept given his hypothesized empirical deadlock, Carnap concludes that the disagreement concerns

> the nature of a linguist's assignment of one of these properties to the predicate as its intension. The assignment may be made explicit by an entry in the German-English dictionary, conjoining the German predicate with an English phrase. The intensionalist thesis in pragmatics, which I am defending, says that the assignment of an intension is an empirical hypothesis which, like any other hypothesis in linguistics, can be tested by observations of language behavior. (1955, 37)

But the term "empirical" in this context is fraught, since in the prior paragraph Carnap has acknowledged that settling that all the facts there are about the extension leaves the intension undetermined. This is why the counterfactual method described there by Carnap and the assumed systematic interrelations remain critical to giving even potential empirical content to the so-called method of intensions.[25]

The point bears emphasizing. Carnap explicitly cites Quine's complaint that in the case just imagined, "the linguist is free to choose any of

---

[25] Recall fn. 2 above.

those properties which fit to the given extension; he may be guided in his choice by a consideration of simplicity, but there is no question of right or wrong" (1955, 37). Against this, Carnap defends the scientific integrity of the method as Carnap imagines it. Carnap's explicit methodological proposal requires using counterfactual cases in order to plumb the intensional boundaries of the concepts in question. Yet Carnap continues not to recognize that Quine objects to an assumption that an underlying system is already connecting dispositions to respond. Absent that assumption, Carnap's investigator cannot possibly achieve what Carnap proposes.

Most importantly for purposes of tracking Quine's unhappiness with Carnap's proposal as formulated here, recall Carnap's previously cited comments regarding "the man on the street" (1955, 40). Carnap's claim is that what such testing elicits from Karl *ex hypothesi* applies to any competent speaker of that natural language. And it could not apply unless some systemic interrelationship already existed. For otherwise it could not be taken that the speakers shared, in the first instance, a language. In this key respect, the "manual" of which Carnap writes *presupposes* what speakers must have in common, which is what it purports to document.[26]

The necessity of assuming a system for constructing K I have elsewhere labeled the "recapture theory" of translation.[27] This view of meaning is explicit in G. E. Moore and reappears in unvarnished form even today.

> Understanding does not presuppose translation, but the other way around. One cannot translate something one does not understand. To translate e_1 from L_1 is to find an expression from L_2 that means the same, and this can be done only by someone who knows what e_1 means …. It is equally absurd to maintain that meaning presupposes translation. Translating an expression

---

[26] This "prior systematicity" assumption emerges in remarks in Carnap's (1942). Although Carnap later distinguishes between what he terms "pragmatics" and "semantics," he ultimately claims that there is a discipline of "pure semantics," one which is "independent of pragmatics" (the study of actual, historical languages and their usages) and that for pure semantics so conceived the choice of rules is "entirely free" (1942, 13). But the situation here is not so clear given Carnap's own account. For antecedent to this characterization of the matter, he writes: "In historical descriptions of particular acts of speaking or writing, expression-events are often dealt with …. When we are not concerned with the history of single acts but with the linguistic description of a certain language or logical (syntactical or semantical) analysis of a certain language system, then the features which we study are common to all events of a design. Therefore, in this kind of investigation, it is convenient to drop reference to expression events entirely and to speak only about designs …. In the same way, if we say in syntax that a certain sentence is provable in a certain calculus, or in semantics that a certain sentence is true, then we mean to attribute these properties to sentence-designs, because they are shared by all sentence-events of a design; the same holds for all other concepts of syntax and semantics" (1942, 6–7). This makes unclear in what sense a choice of rules is "entirely free," inasmuch as while perhaps there can be different sets of rules to the same effect, the underlying "rulishness" of the explananda – the semantic properties – is never questioned.

[27] I first use this phrase in my (1978), 365. I develop it further in my (2019).

e_1 from L_1 by an expression e_2 from L_2 is legitimate only if e_2 means roughly the same in L_2 as e_1 does in L_1. Consequently, the very notion of translation presupposes that the expressions to be translated are meaningful independently of translation, namely by virtue of being used and explained in their home language.[28]

Yet to assert that "the very notion of translation presupposes that the expressions to be translated are meaningful *independently* of translation" presumes an equivalence of meaning existing prior to translation. Translation has a fact of the matter only insofar as it can in principle recapture in one language the meaning present in the other.

Quine consistently maintains that "Factuality, like gravitation and electric charge, is internal to our theory of nature" (1981c, 23). *Meanings, as something to be theorized about, are not in this sense internal to theories of nature, but assumed prior to it.* This point Quine emphasizes in his canonical formulations of indeterminacy. "The question whether, in the situation last described, the foreigner *really* believes A or believes rather B, is a question whose very significance I would put in doubt. This is what I am getting at in arguing the indeterminacy of translation" (1970b, 180–1, emphasis in original). Or again, "We are always ready to wonder about the meaning of a foreigner's remark without reference to any one set of analytical hypotheses, indeed in the absence of any" (1960, 76).

> That it [indeterminacy] requires notice is plainly illustrated by the almost universal belief that the objective references of terms in radically different languages can be objectively compared .... Still one is ready to say of the domestic situation in all positivistic reasonableness that if two speakers match in all dispositions to verbal behavior there is no sense in imagining semantic differences between them. (1960, 79)

In one of his earliest explicit formulations of indeterminacy, Quine also puts the point this way: "What is really involved is difficulty or indeterminacy of correlation" (Quine 1959, 172). The imagined correlation would be between shared structures yielding determinate results – objectively arrived at elements of sets that could then be put into systematic relation.

Quine explicitly acknowledges that seen as purely *internal* matters, truth and meaning appear indistinguishable.

> The indefinability of synonymy by reference to the methodology of analytical hypotheses is formally the same as the indefinability of truth by reference to scientific theories. Also the consequences are parallel. Just as we may

---

[28] Hans-Johann Glock (2003), 204. Livingston offers quotes from others to the same point. See Livingston (2008) 119 and 247, fn. 51.

meaningfully speak of the truth of a sentence only within the terms of some theory or conceptual scheme, so on the whole we may meaningfully speak of interlinguistic synonymy only within the terms of some particular system of analytical hypotheses. (Quine 1959, 170; cp. Quine 1960, 75)

But to conclude from this that a notion of meaning is no worse off than that of truth is, Quine insists, "to misjudge the parallel" (Quine 1960, 75). Because what counts as factuality is arbitrated internal to the theory, determinations of truth remain a theoretical matter. Meaning in the sense that Quine rejects retains an essence because it presumes a noninternal property, or a set of prior even if unknown meaning-making features. "Nonnatural" equates for Quine to something not explainable *within* a theory of nature. As later complaints targeting natural kinds illustrate, a prior structure as it functions for a theory of meaning can be discovered but not accounted for by our theory of the world. That makes it nonnatural, and so marks where the parallel fails.

Paul Livingston aptly notes,

> One significant obstacle, indeed, to understanding the depth and force of Quine's attack against Carnap is that there is great tendency to take the picture of language that Carnap held as inevitable or obviously true. It can seem simply obvious that if speakers share a language, their agreement simply in speaking it must amount to agreement on some corpus of rules, implicit or explicit, in principle capable of formulation and explication. (Livingston 2008, 114)

He later goes on to observe, "The most radical and surprising implication of Quine's indeterminacy thesis is that this assumption of regularity is ungrounded" (Livingston 2008, 122). The account developed in this essay traces exchanges between Quine and Carnap so as to document Quine's evolving realization that his criticisms, starting with those found in "Truth by Convention," leave a theory of meaning without anything to be a theory of.

Although Carnap retreats from attempts to formulate a general theory of meaning, he never surrenders the view that one can articulate a framework of meaning distinct from whatever else it is that one wishes to discourse about. So while Carnap compromises on the scope of a linguistic framework, he retains at the core of his thinking the view that such frameworks make meaning possible. This in turn engenders Quine's successive efforts to reformulate the issue that separates them, efforts which lead to Quine's poignant acknowledgement that "my dissent from Carnap's philosophy of logical truth is hard to state and argue in Carnap's terms" (1963a, 385). Carnap and many who come after him take translation as a recapturing of determinants of meaning. They do not recognize that freed of dogmas – those philosophical pillars that empiricists took to support fixed features that made for meaning – no basis for theorizing meaning remains. Quine ultimately comes to perceive and fully acknowledge a startling conclusion of doubts scouted from early on. His answer to the title question can only be: "Nothing determinate."[29]

---

[29] My thanks to Paul Livingston, James McCord, Sean Morris, Ádám Tuboly, and Thomas Uebel for their comments on earlier drafts of this essay.

CHAPTER 9

Quine and Wittgenstein on the Indeterminacy of Translation

Andrew Lugg

Towards the end of the second chapter of *Word and Object*, W. V. Quine has a footnote commenting on the preceding pages on translation. He writes: 'Perhaps the doctrine of indeterminacy of translation will have little air of paradox for readers familiar with Wittgenstein's latter-day remarks on meaning' (1960, 77, fn.2). This is not a point he had previously made. It is not mentioned in 'Le mythe de la signification', a source of the discussion of translation in *Word and Object*, or in 'Meaning and Translation', a paper adapted from work then being called *Term and Object* (Quine 1958, 4; 1959). Most likely Quine slipped the footnote into the final section of the chapter because he had been alerted — perhaps on one of several occasions he had aired the argument prior to completing the book — to the fact that he was echoing Wittgenstein's post-*Tractatus* thinking. More of an afterthought than a well-flagged observation, the remark is coupled with Wittgenstein's declaration that '[u]nderstanding a sentence means understanding a language', another Wittgensteinian point Quine would have applauded (Wittgenstein 1969a, 5).

Perhaps because the footnote is buried at the end of the chapter and only vaguely gestures at a view of Wittgenstein's, it has mostly been overlooked and its importance as a clue to how Quine intended his discussion of translation to be understood has gone unnoticed. In what follows I explore why Quine might have believed those conversant with the later Wittgenstein's remarks on meaning would readily accept his view of translation. Mainly I hope to show that reading Quine's argument mindful of Wittgenstein's philosophy highlights what he means to establish in Chapter II of *Word and Object* (and, as a bonus, clarifies Wittgenstein's thinking in the *Investigations* about meaning and translation). There can be no disputing that Quine was singularly cool to what he referred to as the 'steadfast laymanship' of Wittgenstein's followers and no denying that Wittgenstein is generally held to stand at the opposite end of the philosophical spectrum from Quine (1960, 262; also 1981a,

70). But they come together – or so I shall argue – on translation and meaning. This does not resolve all the mysteries of Quine's discussion of translation. It does, however, shed light on his argument and dispel more than a few misconceptions about his line of thought. At least on this one matter Quine and Wittgenstein advance consistent rather than competing views.

Quine's aim in Chapter II of *Word and Object* is 'to make [the principle of indeterminacy of translation] plausible' for the special case of '*radical translation*, i.e. translation of the language of a hitherto untouched people' (Quine 1960, 27 and 28). The 'thesis' is that 'manuals for translating one language into another can be set up in divergent ways, all compatible with the totality of speech dispositions, yet incompatible with one another' (Quine 1960, 27; cf. 1992a, 38). At a certain point, Quine would have us notice, the construction of the sought-after translation manual requires 'catapulting oneself into the jungle language by the momentum of the home language' (Quine 1960, 70; 1958, 157).[1] Subsequently Quine modified his account of what is involved in setting up manuals in divergent ways and spoke of the indeterminacy of translation as a plausible hypothesis rather than a provable thesis (Quine 1992a, 47, and Hahn and Schilpp 1998, 728; also Quine 2000, 410, 418, and 419). These are but minor changes, however, and Quine would have continued to think Wittgensteinians would agree that for a manual of translation 'the totality of possible evidence is insufficient to clinch the system uniquely' (Quine 1992a, 101).

While there is no explicit discussion of radical translation in Wittgenstein's published works (or *Nachlass*), he is naturally read as picturing it much as Quine pictured it. Like Quine, who describes himself as tackling 'an anthropological problem', Wittgenstein refers to his 'ethnological approach' (Quine 1937; Wittgenstein 1998, 45).[2] And the thrust, if not the letter, of Wittgenstein's remarks about primitive languages (and language-games) is that a native's speech or text can be rendered, apart from trivial exceptions, in multiple ways. In the *Investigations* when discussing following a rule, Wittgenstein invites the reader to imagine coming into 'an unknown country with a language quite strange to you' and asks: 'In what circumstances would you say that people there gave orders, understood them, obeyed them, rebelled against them, and so on?' (Wittgenstein 1958, §206; also cf. §459). For him, as the surrounding

---

[1] For earlier discussions of radical translation, though not under the name, see Quine (1937), 674, and Quine (1951d), 49ff.
[2] Also compare Rhees (1970), 50: 'This "anthropological" view (*anthropologische Betractungsweise*) is an important feature of the method of discussion or analysis in the *Investigations*.'

text makes clear, it is not logically fixed how a person should proceed, a thought that jibes well with Quine's view that translation is indeterminate.

In any event it is scarcely surprising that Wittgenstein is regularly taken to have nailed his colours to the same mast as Quine when it comes to the indeterminacy of translation. One important commentator on Wittgenstein's later philosophy places him alongside Quine on the indeterminacy of translation since he questions 'whether there are any objective facts as to what we mean' and places an 'emphasis on agreement [that] is obviously congenial to Wittgenstein's view' as he rejects 'any notion that inner "ideas" or "meanings" guide our linguistic behaviour' (Kripke 1982, 55–56). And in a similar spirit a second distinguished commentator writes: 'Both [Quine and Wittgenstein] invoke radical translation, the translation of the language of a wholly alien people, as a heuristic device to illuminate the concepts of language, meaning and understanding … Both recognize a problem of indeterminacy in the use of language and the interpretation of its use' (Hacker 1996, 191; also 218 and 220). While these comments may be faulted for attributing more to Wittgenstein than he commits himself to, they capture the tendency of his thinking well enough (albeit unsupported by textual evidence).

The convergence of Wittgenstein and Quine regarding translation and meaning is actually closer than alleged. It is unclear that 'there are differences' that redound to Wittgenstein's credit because 'Quine sees the philosophy of language within a hypothetical framework of behaviouristic psychology' (Kripke 1982, 56). Whether or not Quine 'would never emphasise introspective thought experiments in the way Wittgenstein does', he is wrongly upbraided for 'bas[ing] his argument from the outset on behaviorist premises', an 'orientation' altogether foreign to Wittgenstein. Nor is Quine at a disadvantage because he rejects 'wholesale … the very concept of meaning' and replaces it with 'a behaviourist ersatz' (Hacker 1996, 203). He was wedded only to the anodyne assumption that in the case of radical translation 'the behaviourist approach is mandatory' (Quine 1992a, 37). While he regarded the concept of meaning as 'ill-suited for use as an instrument of philosophical and scientific clarification and analysis', he reckoned it 'a worthy object of philosophical and scientific clarification and analysis' (Quine 1981c, 185).

The spectre of behaviourism is often raised in discussions of Quine's treatment of indeterminacy and translation, and Wittgenstein is just as often said to have plumped for the behaviourist option. Such interpretations of their remarks are predictable since Quine says: '[T]he thesis of indeterminacy of translation … is a consequence of my behaviourism', and Wittgenstein imagines someone asking: 'Are you not a behaviourist in disguise?' (Quine 1992a, 37; Wittgenstein 1958, §307). Neither philosopher, however, was of the view that – as Wittgenstein puts it – 'everything except human behaviour is a fiction'. What they decry is the idea of an inner world of meanings and the conception of activities rooted in a 'yet uncomprehended process in [a] yet unexplored medium' (Wittgenstein 1958, §308). Both maintain that translation is unsupported by this type of process or medium and take the philosopher's conception of the business as so supported to be 'a grammatical fiction'. Their behaviourism is not a substantive philosophical doctrine. As Quine writes (quoting Paul Ziff): 'Philosophical behaviourism is not a metaphysical theory: it is the denial of a metaphysical theory. Consequently, it asserts nothing' (Quine 1960, 265, fn. 5; also cf. Quine 1969, 296).

But why would Quine think followers of Wittgenstein would respond positively to his remarks on the indeterminacy of translation? Partly this had to be because he believed they regarded language as a social phenomenon. He knew that Wittgenstein stresses the communal nature of language and presumed this to be of a piece with the view that translation is indeterminate. Since Wittgenstein was concerned with 'the spatial and temporal phenomenon of language', not 'some non-spatial, non-temporal phantasm', Quine could have reasonably thought he would agree with the opening remarks of *Word and Object*: 'Language is a social art. Hence there is no justification for collating linguistic meanings [as in translation], unless in terms of man's dispositions to respond overtly to socially observable stimuli' (Wittgenstein 1958, §108; Quine 1960, Preface, first paragraph). Wittgenstein would in all probability have believed, Quine would doubtless have imagined (as he goes on to note), that '[i]n acquiring language we have to depend entirely on intersubjectively available cues as to what to say and when', a consequence of which is 'a certain systematic indeterminacy'. Wittgenstein could not but have supposed, he may well have conjectured, that when one learns language, one depends on 'intersubjectively available cues' and translation is unsecured by 'collat[ed] linguistic meanings'.

Another thought that may have prompted Quine to pen the footnote on page 77 of *Word and Object* is that followers of Wittgenstein would think translation is indeterminate given that they deny the possibility of a logically private language. Nobody who holds there is no such a language, Quine may well have thought, would accept the popular conception of translation of an alien language as a matter of correlating the private goings-on in the native's heads with private goings-on in the linguist's head. In later work – though not in *Word and Object* – the fact that Wittgenstein 'rejected private language' is explicitly paired with the point that it is a mistake to regard 'a man's semantics as somehow determinate in his mind beyond what might be implicit in his dispositions to overt behaviour' (Quine 1969c, 7). Moreover, in still later work, he writes (again with reference to Wittgenstein): 'The question of a private language ... becomes philosophically significant when we recognise that ... language is a social art, socially inculcated', from which, as *Word and Object* has it, the indeterminacy of translation follows (Quine 1981c, 192; see also 46, and Quine 1989, 130).

Quine may also have believed those familiar with the later Wittgenstein's thinking would accept that translation is indeterminate because he regarded them – and Wittgenstein – as linking meaning with use. He would have known Wittgenstein states – discounting names – that 'the meaning of a word is its use in the language' and would have commended his suggestion that words are allied with their use just as money is allied with 'its use', not with what 'you can buy with it' (Wittgenstein 1958, §43 and §120).[3] It is hardly a stretch to envisage Quine thinking, when writing about translation in *Word and Object*, that Wittgenstein was equally of the view that sentences of the target language are correlated with sentences of the home language in as much as they have the same use. Regarding Wittgenstein as a kindred soul about meaning, he would have concluded that he believed that the picture of sentences in different languages as lining up one-to-one has to give way to the picture of them as lining up one-to-many, that is, that the conception of translation as determinate has to give way to the conception of it as indeterminate.

Whatever prompted Quine to refer to Wittgenstein on meaning at the end of Chapter II of *Word and Object*, he was not wrong in thinking Wittgensteinians take language to be a social affair, are antipathetic to the possibility of a private language and equate meaning with use, all of which he takes to sit uncomfortably with regarding translation as determinate. What he apparently believed he had contributed to the discussion was not so much the thought that translation is indeterminate but an argument for concluding that it is. There is in any case no comparable detailed defence in Wittgenstein's writings of the view that 'the radical translator is bound to impose about as much as he discovers' (Quine 1992a, 49). Much less clear is how Quine understood the indeterminacy

---

[3] Compare Quine (1981c), 46: 'Wittgenstein has stressed that the meaning of a word is to be sought in its use'; also 192 on the idea that 'a legitimate theory of meaning must be a theory of the use of language'.

of translation and how Wittgenstein would have understood it had it been put to him. To hold that translation is indeterminate is to hold that it is not determinate, but what is it about viewing it as determinate that Quine – and I am suggesting Wittgenstein – deemed objectionable? Neither philosopher was simply retailing 'the platitude that uniqueness of translation is absurd' (Quine 1960, 73). But if not this, then what exactly were they jointly protesting?

It is often missed by those outside the circle of close readers of Quine that he does not regard translation ordinarily understood and common-or-garden conceptions of meaning as suspect. It is not true, as one influential commentator has it, that Quine 'give[s] up on meaning and translation entirely' (Soames 2003, 377). Rather the reverse, he writes: 'I am strong for translation, in which I have been much involved in a practical way' and reminds a critic that he is 'in favour ... of translation, even radical translation' (Hahn and Schilpp 1998, 73; also 728; Quine 1969e, 312). And Wittgenstein, likewise, was comfortable with translation, typically pursued, and everyday conceptions of meaning. He encouraged translation (of part) of *The Brown Book* into German and did not hesitate to suggest improvements to translations of the *Tractatus* and the pre-war version of the *Investigations*.

Though less obviously wrong, it is as much a mistake to suppose Quine and Wittgenstein regarded translation as indeterminate because they took language to be shaped by cognition and believed the systems of concepts of speakers of alien languages differ from our system of concepts. While Quine flirted for a time with the Sapir–Whorf hypothesis about the influence of language on thought, he did not invoke the hypothesis in *Word and Object* (Quine 1980, 61). In fact when discussing the indeterminacy of translation in Chapter II he puts distance between his thinking and Whorf's (Quine 1960, 77–78; also Quine 1959, 171–172). Nor is there passable textual evidence for aligning Wittgenstein with Sapir and Whorf. It is of no relevance that he was of the opinion that 'to imagine a language is to imagine a form of life' and says: 'If a lion could talk we could not understand him' (Wittgenstein 1958, §19 and 223). To the contrary he speaks of definitions as 'rules for the translation of one language into another' and takes the trouble to specify an essential condition for 'translat[ing] some otherwise alien form of expression into ... our customary form' (Wittgenstein 1922/1990, 3.343; 1958, 175).

Quine – and, I believe, Wittgenstein in his own way – was taking issue with the conception of translation promoted by philosophers and philosophically minded non-philosophers, the central pillar of which is that, ignoring practical difficulties, there is a single acceptable way of lining up sentences of a target language with sentences of the home language. Quine is not, as claimed by the editor of the volume in which 'Meaning and Translation' appears, addressing 'the student of foreign languages [and] the anthropologist attempting to understand and describe an alien culture' (Brower 1959, 7). Rather he is, as he says in *Word and Object*, making a 'philosophical point' (Quine 1960, 29; also 1969c, 34, and 2008a, 345). And similarly Wittgenstein believed philosophers are bamboozled when they construe translations as uniquely capturing the meaning of a sentence or text. The object of both philosophers' criticism is the uniqueness of good translations, not their existence. It was Quine, but it could have been Wittgenstein, who said: 'Of course translation must go on. Indeterminacy means that there is more than one way; we can still proceed to develop one of them, as good as any' (Quine 2008a, 251).

What Quine and Wittgenstein jointly repudiate is the simple but exceedingly inviting view that in translation something is passed along from one language to another and that translations are acceptable just to the extent that this occurs. The assumption is that in good translation the meanings of sentences of the target language are returned as meanings of sentences of the home language. The reason that 'The sky is blue' is held to translate 'Le ciel est bleu' and 'Der Himmel ist blau', for instance, is that it picks out the same idea, thought, meaning that they pick out. However different the target and home languages, there is a neutral touchstone against which translations are properly assessed, an impartial criterion that determines whether sentences of the two languages match. In other words, determinate translation is anchored translation, indeterminate translation unanchored translation. The reason that 'The sky is blue', 'Le ciel est bleu' and 'Der Himmel ist blau' are interchangeable is that they have the same use. Designating, labelling, picking out the same meanings does not come into it.

It is with an eye to supporting this conclusion that Quine and implicitly Wittgenstein consider the task of rendering 'the language of a hitherto untouched people' when 'all help of interpreters is excluded' (Quine 1960, 28). As I read them, they take the extreme case of radical translation to dispose of the philosophical conception of translation as determinate. For both the discussion of radical translation is, as Quine came to refer to it, a 'thought experiment' (Quine 2000, 419; 2008a, 284 and 341; and 1992a, 38). Simply stated, their claim is that consideration of the task of penetrating an alien language just as definitively shows that translation is not determinate as consideration of how light and heavy objects dropped from a tower behave shows their rate of descent is not different. (Another example, perhaps closer to Quine's thinking, would be Einstein's thought experiment concerning the simultaneous occurrence of events.)

When Quine is understood as arguing against the philosopher's conception of determinate translation, it is unsurprising that he should mention Wittgenstein's repudiation of the notion of a logically private language, stress that language is a social art and recommend concentrating on the uses of words instead of their meanings. He believed that denying the existence of a logically private language goes hand in hand with denying that translation is secured by mental entities (and rejecting the assumption that 'The sky is blue' translates 'Le ciel est bleu' because the same mental idea, representation or the like is in the heads of English speakers as in the heads of French speakers). Furthermore, he – and arguably Wittgenstein – believed that recognising that language is a social art and 'meaning [is] use' is antithetical to picturing translation as ensured by something attached to (or beyond) language, something that objectively moors it and renders it determinate. In the end it was, I fancy, this – along with the thought experiment of Chapter II of *Word and Object* – that prompted Quine to believe he and Wittgenstein agreed that translation is indeterminate and to state that the doctrine would have 'little air of paradox' for those conversant with Wittgenstein's later view of meaning.

The first paragraph of 'Meaning and Translation' and the first paragraph of Chapter II of *Word and Object* provide yet more insight into Quine's view of the indeterminacy of translation. In the paper he announces that he is concerned with the question of 'what remains when, given discourse together with all its stimulatory conditions, we peel away the verbiage', a question he plans to explore by considering how a linguist would 'penetrate and translate a hitherto unknown language', it being 'a good way to give a meaning to say something in the home language that has it' (Quine 1959, 148). And in the book he begins by noting that 'it is to [the present and past barrage of non-verbal stimulation] that we must look for whatever empirical content there may be [in our voluminous and intricately structured talk]' (Quine 1960, 26). What he would like to show is that two persons 'alike in all their dispositions to verbal behaviour' would differ in identical conditions in their 'meanings or ideas' but, this way of stating the point being questionable, he switches to a consideration of translation. In both works his primary aim is to convince the reader that whereas '[o]bservation sentences peel nicely', other sentences 'lack linguistically neutral meaning' (Quine 1959, 171; 1960, 76–77).

Quine could hardly state it more candidly that he is directing his fire at the philosopher's conception of (meaningful) sentences as accompanied by propositions, meanings, thoughts, ideas or the like. He maintains that translation is wrongly regarded as determinate because there is nothing of this sort accompanying sentences and hence nothing that is passed along when sentences of one language are rendered as sentences of another language. What comes under the hammer is '[u]ncritical semantics', the 'copy theory of language', 'the myth of the museum in which the exhibits are meanings and the words are labels' (Quine 1969c, 27). This myth, along with the associated idea that '[t]o switch languages is to switch labels', has to go, a conclusion no reader familiar with 'Wittgenstein's latter-day remarks on meaning' will balk at. From beginning to end Quine deplored the philosopher's conception of meaning, and the later Wittgenstein, working on a parallel track, found it equally dreadful.[4]

Should there be any lingering doubt that Quine is as concerned with meaning as with translation, it is worth recalling that Chapter II of *Word and Object* is titled 'Translation and Meaning' and that he writes: '[T]he discontinuity of radical translation tries our meanings: really sets them over against their verbal embodiments, or more typically, finds nothing there' (Quine 1960, 76; 1959, 171). Indeed he later says: 'I don't recognise a problem of indeterminacy of translation. I have a thesis of indeterminacy of translation, and its motivation was to undermine Frege's notion of a proposition or *Gedanke*' (Quine 1990b, 176). Moreover, should there be any doubt that this was his motive, it should be noted that he also writes: 'What was challenged was the philosophical notion of propositions, the meanings of sentences' and states that the thought experiment in radical translation 'was meant as a challenge to the reality of propositions as meanings of cognitive sentences' (Quine 2000, 410 and 419; also cf. 418). For him radical translation is indeterminate, not 'because the meanings of sentences are elusive or inscrutable', but 'because there is nothing to them, beyond what [the radical translator's] fumbling procedures can come up with' (Quine 1992a, 47; also cf. 1969c, 5; 1969e, 304; and Hahn and Schilpp 1998, 367).

If there is anything certain about Wittgenstein's thinking, it is that he was as opposed as Quine to equating propositions with meanings understood as what sentences of different languages alike in meaning share. He too considered it a philosopher's myth that sentences ordinarily understood are correlated with or accompanied by mental entities, abstract

---

[4] Also compare Hahn and Schilpp (1998), 74: 'It is Wittgenstein's rejection of a private language. It is this, and not mentality as such that disqualifies any irreducibly intuitive notion of meaning or synonymy or semantic relevance.'

Platonic meanings or other non-linguistic objects. In his view the closer the functioning of language in everyday life is examined, the harder it is to embrace this philosophical conception and to agree that a successful account of translation must appeal to propositions, meanings or thoughts as language-independent contents of declarative sentences. Thus we find him disparaging the notion of '[m]eaning-body [*Bedeutungskörper*]' and, focusing on the special case of words in sentences, saying: '[T]here isn't anything hidden ... The function must come out in operating with the word' (Wittgenstein 1958, §559). Small wonder that he should say when discussing the question of what a person in a strange land means: 'The common behaviour of mankind is the system of reference by means of which we interpret an unknown language' (Wittgenstein 1958, §206).

Lest it be supposed that Quine and Wittgenstein are tilting at windmills, it deserves reiterating that they oppose the seductive idea of sentences as accompanied by meanings, not just the narrower notion of meanings as things. In the discussion following his presentation of 'Le mythe de la signification', Quine dismisses the complaint that he treats meaning as an entity comparable to the nineteenth-century economist's 'Buying Power' and overlooks the idea of meaning as a disposition comparable to 'buying power' (Quine 1958, 180). He has, he says, no 'special objection' to meanings as entities – even to 'Buying Power' – his sights being set on the notion of 'meaning as such' (Quine 1958, 182 and 139). Moreover, in a similar vein Wittgenstein is opposed to meanings of words and sentences whether understood as mental, Platonic, physical or dispositional entities, these all being, he believed, out of place in serious philosophy. It is not by chance that in the discussion in *The Blue Book* referred to in footnote 2 on page 77 of *Word and Object* Wittgenstein targets the thought that when 'looking for the use of a sign ... we look for it as though it were an object *co-existing* with the sign' (Wittgenstein 1969a, 5).

The argument regarding translation which Quine advances in Chapter II of *Word and Object* (and which Wittgenstein would, as I read him, have welcomed) is not complicated. It turns on the assumption that translation would be determinate were there language-independent meanings and were the interchangeability of sentences in different languages objectively determinable. Given this assumption, it follows that alternative manuals of translation are never equally good, there being an independent way of telling whether sentences mean the same. Hence, since alternative manuals are invariably possible, Quine argues, translation is not determinate and it is a mistake to assume the existence of propositions, thoughts or other language-independent meanings attached to sentences. The claim is not that translation is indeterminate since there are no propositions but rather that there are no propositions since translation is indeterminate. While Quine is on record as holding that propositions are unacceptable since it makes no sense to speak of them as having beginnings and ends, in Chapter II he proceeds from different premises (Compare Quine 1960, 200ff). The indeterminacy of translation, as its name indicates, has to do with a feature of translation that philosophers encumber it with, namely that its success or failure depends on the propositions undergirding the sentences of the languages in question.

If not fairly standard, this conception of translation is far from outmoded. More than a few philosophers nowadays picture words and sentences as having attached meanings and many more regard Fregean *Gedanken* and Russellian propositions as philosophically indispensable. Quine's 1934 criticism of ontologically robust conceptions of the propositional calculus continues to have purchase, as does Wittgenstein's view that commonplace thinking about the equivalence of 'p' and 'not-not-p' is plagued by the 'mythology of symbolism' (Quine 1976, 265–271; Wittgenstein 1974, 53).[5] Nor should it be forgotten that cognitive scientists as well as philosophers of language take the existence of mental phenomena in the form of representations pretty much for granted. Quine undoubtedly invites misunderstanding when he refers to translation as indeterminate rather than states outright that he is mounting an argument against propositions. But this is what he argues and had he said so, he would have received less criticism, not to mention less attention. Nor would he have been repeatedly obliged to emphasise that he was not claiming translation is problematic.

Recognising propositions, meanings, beliefs-in-the-head and such like is tantamount in the eyes of both Quine and Wittgenstein to recognising a place for synthetic a priori knowledge and truth, something that neither had any time for. It was central to the thinking of the one no less than the other that earlier philosophers mistakenly regard philosophy as a subject that combines the necessity of logic with the fact-based character of science and fail to appreciate that respectable inquiry is either logical or scientific. From first to last Wittgenstein maintained that there is no enterprise between logic and science, and that information about the world is obtainable only by empirical investigation (Wittgenstein 1922/1990, 6.3; Wittgenstein 1977, II.3). And Quine was just as adamantly of the view that there is no in-between subject and equally opposed to past philosophical speculation, the results of which he unceremoniously consigned to 'the

---

[5] Also see Wittgenstein (1958), 147, fn. (a); and Wittgenstein (1979), 4 and 50–51.

abyss of the transcendental' (Quine 1981c, 23). It is beside the point that Wittgenstein took logic and mathematics to be sharply separated from science while Quine took them all, with the possible exception of the higher reaches of mathematics, to form a single department.

Quine had no serious quarrel with the account of '[t]he right method of philosophy', as Wittgenstein characterised it. He too believed (logic and mathematics aside) that philosophers should say 'nothing except what can be said' – that is, only empirical propositions – and was no less firmly of the opinion that the path of clarification and criticism is the sole other path open to sound philosophising (Wittgenstein 1922/1990, 6.53). For him, as for Wittgenstein, philosophers who treat translation in terms of lining up propositions are saying something metaphysical and are subject to the charge of having 'given no meaning to certain signs in [their] propositions'. He agrees with Wittgenstein on what counts as the 'the only strictly correct method' but applies 'philosophy' more broadly. While agreeing that scientific problems have 'nothing to do with philosophy' in Wittgenstein's sense, he felt a kinship with old-style natural philosophy and its goal of a comprehensive theory of the world (Quine 1981c, 191). The difference in what a word covers is, however, not something he or Wittgenstein was disposed to debate (Quine 2000, 411; Wittgenstein 2005, ch. 89).

When criticising traditional philosophical thinking about propositions, meanings and translation, Wittgenstein and Quine use different language but, practically speaking, come to the same conclusion. Wittgenstein holds that '[m]ost propositions and questions, that have been written about philosophical matters, are not false but senseless' and would have regarded philosophers who regard translation as a matter of passing along propositions or meanings as mystery-mongering (Wittgenstein 1922/1990, 4.003). And while holding that many remarks that Wittgenstein considers nonsensical to be trivially false, Quine has no compunction about dismissing past philosophy as meaningless (Quine 1960, 229). He treats meaninglessness and nonsense as useful terms of criticism and differs from Wittgenstein more in emphasis than strict doctrine. Indeed, at the beginning of Chapter II of *Word and Object* he states that speaking of a person's meanings or ideas 'invites the charge of meaninglessness' and plainly thinks that claiming propositions guarantee the determinacy of translation is worse than claiming phlogiston is released during combustion (Quine 1960, 26).

Quine may have preferred to speak of facts, Wittgenstein to speak of concepts, but they understand radical translation in essentially the same way. Quine's thought experiment and Wittgenstein's example of translation of an alien language are comparable to Galileo's discussion of how heavy objects fall and Einstein's discussion of simultaneity, discussions frequently and reasonably regarded as philosophical. Moreover, when it suited his purpose, Quine made use of conceptual analysis, even appealed to everyday usage (cf. Quine 1992a, 2; and 1976, 229). And Wittgenstein for his part did not think twice about citing empirical facts when he considered them pertinent (see, e.g., Wittgenstein 1977, III.28, on perfect pitch and III.31 on color blindness). What matters is how Quine and Wittgenstein treat the subject, not their general pronouncements about philosophy and how they describe their respective 'methods'. As Wittgenstein noted in another context, it does not matter what one chooses to say, 'so long as it does not prevent [one] from seeing the facts' (Wittgenstein 1958, §79).

Besides spurning Russellian propositions and Fregean thoughts and dispensing with normal philosophical thinking about translation, Wittgenstein and Quine are at pains to separate science from metaphysics and distinguish 'the propositions of natural science' from the synthetic a priori propositions ubiquitous in the philosophical literature. Contrary to a widespread misconception, Wittgenstein was not anti-science, only an uncompromising opponent of scientism. He holds, indeed insists, that science 'can be said' and recognises the possibility of recasting – and evaluating – at least some metaphysical propositions as scientific propositions (cf. Wittgenstein 1922/1990, 4.1121). And Quine, still more clearly, promotes science as an alternative to traditional speculative philosophy and allows for the possibility of reformulating philosophical concepts, ideas and problems in some instances as scientific concepts, ideas and problems. Thus, he would 'rescue' from 'the abyss of the transcendent' the question of 'the reality of the external world', radical scepticism, the possibility of a 'rational reconstruction of the world from sense data' and the philosopher's notion of 'a matter of fact' (Quine 1981c, 22–23).

Where Quine and Wittgenstein differ is not on the question of 'the only strictly correct method' but on the very different question of which way to go on. In the *Tractatus* Wittgenstein identifies two paths for philosophers who scorn metaphysical thinking (and synthetic a priori knowledge and truth), the critical path that he pursues and the scientific path that Quine pursues. Wittgenstein focuses on what he deems unbridled philosophical theorising and limits himself to examining and criticising the terms in which philosophers concoct their explanations and theories, while Quine takes on the mantle of science and offers scientific answers to philosophical questions he believes can be profitably reformulated as scientific questions. What Wittgenstein sees as philosophy enough, the criticism of past philosophy, Quine sees as a prolegomenon to what interests him most, the development of a scientific theory of how we come to have our theory of science. He does not duck the task of criticising past philosophy – Chapter II of *Word and Object* is ample evidence of that – but he dispatches it as quickly as he can in order that – as in he says when musing on the notion of 'mile', a rumination highly reminiscent of Wittgenstein's handling of philosophical confusion – 'we can get on' (Quine 1960, 272).

On the view I am defending, Quine and Wittgenstein are to be understood as responding in different but compatible ways to the demise of the philosophical picture of sentences as accompanied by meanings and translation as involving the passing on of meanings. Wittgenstein sticks to philosophy narrowly construed and bends his energies to combatting philosophical thinking about propositions, while Quine happily trades in philosophy for science and seeks to contribute to our knowledge of the world. Wittgenstein is unconcerned with questions that fall in the province of science, the questions Quine takes to be all-important included. He would not dismiss but had no interest in the project that Quine worked on for more than four decades, that of providing a 'rational reconstruction of the individual's and/or the race's actual acquisition of a responsible theory of the external world', one that makes no appeal to propositions, meanings or other scientifically unacceptable notions (Quine 1995a, 16).

At the risk of unnecessarily repeating myself, I would underline that in Chapter II of *Word and Object* Quine is engaged in a twofold exercise of criticising the philosopher's notions of proposition and developing an account of our acquisition of language and our theory of the world free of the notion. He was reluctant to present negative criticism without an accompanying positive theory, and in this chapter he develops his argument against propositions in tandem with remarks about how science is generated from stimulation. Still there are two projects, not one, the criticism of what he takes to be philosophical gobbledygook being orthogonal (and prior) to his on-going project of explaining how we manage to acquire descriptive language (and our conceptual scheme). The reason the chapter is as long as it is, in fact the longest in the book, clocking in at some fifty-four pages, is that it includes remarks on stimulus meaning, occasion and observation sentences, the synonymy of terms, the translation of the logical connectives, analyticity and various other topics that bear only marginally on the theme of the indeterminacy of translation. As Quine himself says: '[T]he chapter [runs] longer than it would if various of the concepts and considerations ancillary to [the job of making plausible the indeterminacy of translation] did not seem worthy of treatment also on their own account' (Quine 1960, 27).

The suggestion being canvassed here, then, is that Quine and Wittgenstein differ regarding the indeterminacy of translation – and arguably much else as well – in attitude rather than belief. They agree on a great deal and share to a considerable extent a similar philosophical background. But they are temperamentally unlike. Each is motivated by his own predilections and approaches the question of meaning and translation from his own angle. There is no substantial conflict between concentrating on intellectual sleight-of-hand, as Wittgenstein does, and pursuing scientific questions in a scientific spirit, only a difference of preference. The two projects are complementary, not opposed, and are most helpfully regarded as alternative ways of proceeding from a joint starting point. Quine would never say, as Wittgenstein says: 'Scientific questions may interest me, but they never really grip me. Only *conceptual* & *aesthetic* questions have that effect on me. At bottom it leaves me cold whether scientific problems are solved; but not those other questions' (Wittgenstein 1998, 91). Rather the opposite. Scientific questions gripped Quine and while conceptual questions did not leave him completely cold, he took them to be unavoidable diversions.

There are, it seems to me, considerable advantages to reading Quine and Wittgenstein as proceeding on parallel paths. Whether all their reflections admit of the kind of treatment I have argued is appropriate for the question of the determinacy of translation, it is a good working hypothesis that Quine intends to contribute to science as opposed to philosophy as traditionally construed, Wittgenstein to criticise philosophical thought as opposed to advancing a view of the world, language, the mind or anything else. For instance, their seemingly differing views regarding epistemology and their apparently opposed conceptions of language acquisition are easily reconciled when Quine's scientific project is separated from Wittgenstein's critical project (compare Wittgenstein 1922/1990, 4.111, and 1958, §§28–30, with Quine 1969c, 31, and 2008a, 467). In addition, reading Quine and Wittgenstein as suggested shows Wittgenstein to be wrongly treated as an ordinary language philosopher, Quine wrongly treated as an ordinary science philosopher. Both are reorienting the subject. Quine points out that his conception of the subject is '[a] far cry ... from old epistemology', and Wittgenstein, doubtless overenthusiastically, sees himself as having inaugurated 'in philosophy, a "kink" in the "development of human thought"' and 'a "new method" had been discovered, as had happened when "chemistry was developed out of alchemy"' (Quine 1974a, 3; Moore 1993, 113, quoting Wittgenstein).

Afterword: Carnap, Quine and Wittgenstein on the indeterminacy of translation

Rudolf Carnap is naturally pictured as a swing figure between Wittgenstein and Quine, not least on the topic of the indeterminacy of translation. He was influenced in no small way by the Wittgenstein of the *Tractatus*, and Quine was importantly influenced by his *Logical Structure of the World* and *The Logical Syntax of Language*. But it is not obvious how far he goes along with them on translation. While he was as much against metaphysics and other sorts of philosophical speculation as Wittgenstein and Quine and just as ill-disposed to the possibility of synthetic a priori knowledge, he cannot be said to have followed faithfully in Wittgenstein's footsteps or to have wholeheartedly embraced Quine's discussion. Nor is it as clear in his case as it is in Quine's and Wittgenstein's how he stood regarding 'the only strictly correct method' (as the *Tractatus* has it). Unlike Wittgenstein he was loath to ban the philosophical treatment of language and unlike Quine he was loath to trade philosophy for natural science.

Carnap was aware of Quine's discussion of the indeterminacy of translation in Chapter II of *Word and Object* but does not directly engage with it. There is, however, reason to think he would have accepted, if not welcomed, Quine's argument. He commented in the margins of the chapters following Chapter II of his copy of *Word and Object* but let this chapter (and Chapter I) pass unmarked (Creath 1994, 292). And, as has been also noted, in *Meaning and Necessity* (1947) he rehearses an argument similar to the argument Quine rehearses in *Word and Object*. 'A predicator in a word language (e.g. "gross" in German) or in a symbolic language (e.g. an abstract expression in Quine's system) may', he says, 'be regarded as the name of a class but also as the name of a property', an observation congenial to Quine's thinking about translating words of a wholly alien language (Carnap 1947/1956, 100). It is immaterial that in later work Quine distinguishes between the indeterminacy of translation as applying to sentences and as applying to terms (e.g., Quine 1992a, 50). In *Word and Object* he treats them together.

Had Carnap realised or been told that Quine intended the argument of Chapter II of *Word and Object* as 'a challenge ... to the reification of meaning, notably propositions', he would not have found the argument of the chapter any less acceptable (Quine 2000, 418). He was not more sympathetic than Quine (and Wittgenstein) to the idea that sentences are accompanied by meanings (though there are remarks in his later work that lean this way). But in 'Meaning and Synonymy in Natural

Languages' (1955), he responds to Quine's animadversions about the philosopher's notion of meaning by outlining what he takes to be a philosophically respectable way of determining the meaning of a word (Carnap 1947/1956, 236–240). Such a tactic is not one Quine would find unacceptable (apart from the fact that the account of meaning Carnap provides is language-specific rather than, as Quine demands, language-general). Quine had nothing against an empirical treatment of meaning, in fact offers what he variously calls a 'behavioristic ersatz' and a 'vegetarian imitation' (Quine 1960, 66 and 67). For him, this ersatz is a scientific concept belonging to the subject he takes to supersede traditional philosophy and it is senseless to hanker for more, nothing more being needed.

Still Carnap's official view is close to the view pioneered by Wittgenstein. He too thinks it falls to philosophers to analyse concepts and in an unmistakable echo of Wittgenstein's equation of philosophy with the logic of our language, he equates philosophy with the logic of scientific language (Wittgenstein 1922/1990, 4.0031; Carnap 2002 [1937], xiii). And it cannot be accidental that Carnap found nothing especially untoward about Quine's discussion of the indeterminacy of translation, the critical remarks of Chapter II of *Word and Object* being in his view properly philosophical (and the theoretical observations in the chapter being, in intention at least, properly science). It should not be forgotten, however, that Carnap differed from Wittgenstein in believing it possible and necessary to accord philosophy a much wider scope. Whereas Wittgenstein believed different kinds of concepts were subject to different kinds of critical analysis, Carnap sought to firm up Wittgenstein's thinking, provide a philosophically legitimate account of meaning and hold on to a hard-and-fast distinction between the deliverances of philosophy and the deliverances of science. This left him open to Quine's continuing criticism, criticism that Wittgenstein with his uncompromisingly critical conception of philosophy managed to escape.[6]

---

[6] I am grateful to Paul Forster for detailed comments on an early draft, to Gary Kemp for pulling me up at several important points, to Bela Szabados for checking the final version of the essay, and to Sean Morris for no little editorial assistance. I think it was Peter Hylton who drew my attention to Quine's 1937 paper 'Is Logic a Matter of Words?', and I have Douglas Quine to thank for permission to cite it. In conversation some twenty years ago Burton Dreben impressed on me that Quine and Wittgenstein are not as different as often assumed. Whether he would approve of the argument in this essay is another matter.

CHAPTER 10

Turning Point
Quine's Indeterminacy of Translation at Middle Age

Richard Creath

My unofficial topic here is middle age. (Whose? We'll see in a moment.) But the first thing I want to do is to quote that famous American philosopher Doris Day, who said, "The really frightening thing about middle age is that you know you'll grow out of it." I'm not sure why that should be so frightening – when you consider the alternative. Worse, there are those who never grow into it, who never really grow and develop at all. Of course, interesting lives are the ones that do grow, the ones that have a beginning, a middle, and an end. The same goes for philosophic lives, or so I am inclined to believe.

Now my official topic is *Word and Object*, Quine's masterpiece, published in 1960. More precisely I am concerned with Quine's arguments there concerning the indeterminacy of translation. Quine was fifty-two when the book was published, middle aged by most reckoning. This time of life is often a time of reflection and review and possibly for a mid-course correction. And that is the sort of turning point that *Word and Object* shows. It is no mid-life crisis. So my thesis is that Quine's arguments constitute an important turning point in the development of Quine's philosophy. The philosophic interest that I would claim for this thesis lies not in the fact that *Word and Object* is such a turning point, but rather in the how and the why of it.

Unsurprisingly, the first obstacle to my thesis is the widely held idea that Quine's philosophy doesn't have turning points, that he has a view but not a history. He may have added to his view over the years, and he certainly filled in details, or even changed very minor points. But the essentials are all right there in his earliest works. Now I think that Quine is too interesting to suppose that this even could be right; I see many twists and turns. But there are at least two causes of the prevailing tendency to see Quine's work as a largely unchanging whole. First is the astounding success of "Two Dogmas of Empiricism" (Quine 1951e). It has been translated into dozens of languages and been reprinted countless times. It may be the only work of twentieth-century philosophy that every American philosophy major reads, and for many, if not most, it is the only work by Quine that they will ever read. For them, Quine *is* "Two Dogmas." And you can't have much growth and development within the confines of a single paper. Moreover, in teaching Quine to students, it is hard enough to present a snapshot of his wonderfully subtle views. To capture the full video of his evolving opinions is well beyond the time constraints of most courses. The second cause of the mostly-changeless-Quine view is Quine himself. He tended to read his earlier works in the light of the later ones, and if necessary to read into them his later ideas. Moreover, he often liked to think that the later work added to but did not significantly alter his previous writing. So he sometimes reaffirmed his earlier work even when his own writing had superseded it. Given the subtlety and difficulty of his thought and the plausible idea that surely Quine must be the best guide to his work, it is hardly surprising that much of the philosophic world takes Quine's no-major-changes view seriously.

Well, Quine is, sadly, no longer with us. So we will have to think this through for ourselves. And *Word and Object* is itself substantially more than fifty years old, middle aged by most standards, at least for humans anyway. But what could it even mean to say that a book is middle aged? Philosophical books of this caliber are immortal. But the tradition of its interpretation is also as old as the book. And that tradition is, I think, in need of a mid-course correction. Perhaps such traditions always are. If we are here to effect such a correction in our understanding of the book, it is not enough to suppose or assert that there were turning points. We will have to trace the changes and underscore their importance. So my plan is this. First, I highlight some features of Quine's pre-*Word and Object* view. These both motivate some of the views expressed in the 1960 book and provide benchmarks against which to measure change. Second, I examine the two arguments for the indeterminacy of translation. They have different strengths, but each has its role to play. Third and finally, I look at some features of Quine's views as they emerged after *Word and Object*. This will show some of the changes in Quine's philosophy and link those changes to the arguments for indeterminacy.

10.1   Classic (Pre-*Word and Object*) Quine

There are two aspects of Quine's view in the 1950s that I want to emphasize here. One is semantic, the other epistemic. The semantic point may seem too obvious to mention, but I will need it later on. And that point is that

Quine is a thoroughgoing extensionalist. By 1950 he had come to reject as unintelligible all intensional notions such as necessity, synonymy, meaning, and most notoriously, analyticity. These terms might be mutually interdefinable, but this is not enough to make them meaningful. What is required for their salvation is the availability of behavioral or other empirical criteria so that they can be used in a real empirical theory of language. In short, what Quine is invoking is a rather unspecified version of an empiricist criterion of significance, just like the ones listed by Hempel in his famous papers (1950, 1951) of that era. Some have interpreted Quine as demanding "explanatory" or "principled" concepts. Such interpretations are often no clearer than Quine's own talk of behavioral criteria. While not wrong, these interpretations really don't add much specificity to the empirical criterion of significance already at hand.

As for Quine's not specifying the criterion, that probably adds to the rhetorical strength of his claim. An undefined accusation, like an unnamed threat, is always more menacing because it is unclear how to defend oneself. Besides, trying to lay out a viable empiricist criterion of meaningfulness was already proving to be a minefield that was prudent to avoid. Beyond this, there is, for present purposes, no need for us to assess the legitimacy of Quine's demand or the prospects for meeting it. For now, it is enough to note that in the 1950s Quine rejects intensional notions and why.

The flip side of the rejection of analyticity etc. is, unsurprisingly, a focus on the interrelated concepts of truth, reference, and ontology. Such notions he thought to be clearly and straightforwardly understood. This asymmetry between Quine's treatments of reference and of meaning is perhaps the most salient feature of his views of the 1940s and 50s. Even without a definition of "true-in-L" for variable L, we are told (Quine 1953b), the attribution of truth to "snow is white" is as clear to us as the attribution of whiteness to snow. While there was no attempt to show that "refers to" meets an empiricist criterion of meaningfulness, neither Carnap nor anyone else had denied that this term was intelligible. Presumably we are free to take it as a primitive and to build our account of language thereon. And that is just what Quine did. Quine also explored the related notions of ontological commitment and ontological reduction, giving a criterion for the former (Quine 1948a). Ontology was central to Quine's concerns before 1960. His preferred ontology was physicalist and included both middle-sized physical objects of everyday experience and also the unobservable entities of theoretical physics. We'll return to this ontology in a bit.

For now, it suffices to repeat that prior to *Word and Object* Quine rejected analyticity etc. as unintelligible and focused his semantic accounts on what he considered the fully intelligible notions of reference and ontology. There is also a second aspect of Quine's work from the 1950s that I want to highlight here, namely his epistemic concerns. The first thing to note is how sketchy Quine's concern with epistemology is prior to *Word and Object*. Remember that Quine had been publishing in philosophy for nearly thirty years by that point, but the last section of "Two Dogmas" is about all there is on the subject. While what he says there is fascinating and important, it is also very sketchy. In addition to "Two Dogmas," there are also a few pages in the "Introduction" to *Methods of Logic*, published the previous year. Those pages, however, cover virtually the same ground in virtually the same terms as "Two Dogmas" and so add only a bit more emphasis on the role of logic and mathematics. Beyond this there are only a few scattered remarks, and taken all together this is not much attention to epistemology over nearly three decades.

Even in that sketch, however, there are a few things to note:

1. It is a one-tier system, that is, all our beliefs meet experience together.
2. Underdetermination is a major theme and commitment.
3. Simplicity plays an important testing role in Quine's account, even as a kind of evidence.

Let's look at these three in turn. The most striking thing about Quine's sketch at the end of "Two Dogmas" is that it is a one-tier system. All of our beliefs meet experience together. Logic and physics and the most casual matters of history are in the same boat – Neurath's. They meet experience together and in the same way.

Then there is underdetermination, a major commitment. Unfortunately, what that view is itself underdetermined by what he says. I doubt if anyone can say for sure what Quine's view of underdetermination is at this period, but we can make an educated judgment. What are the possibilities? One is that statements describing our observational evidence by themselves do not entail (via elementary logic) any one of the many theoretical hypotheses that compete in a given domain. Nor do they rule out all but one. Another possibility is that underdetermination is meant to convey that these observational statements together with elementary logic and such extra-evidential principles as simplicity and conservatism do not pick out one of the theoretical hypotheses as superior to or likelier than the alternatives. Even these two possibilities have variations (which we will largely ignore) depending on whether we are to take the observational statements as those we are in a position to make right now, as all those we will ever be in a position to make, or as all those truths of the appropriately observational sort whether anyone is in a position to make them or not.

In addition to these two general possibilities, one involving principles like simplicity and the other not, there is also a psychological version of underdetermination. To choose a very simple case, the quantified claim 'there are white swans' is logically determined by the combination of the observation sentence 'this swan is white' and elementary logic. But there could be someone so stubborn or so unwilling to have any commerce with quantifiers as to refuse to believe the quantified claim. The idea that human beings are sometimes stubborn, willful creatures is not new, sanctioned as it is by ancient religious tradition. Quine as a general rule tends to avoid reinforcing religious dogma, so perhaps one ought to be cautious before attributing such a version of underdetermination to him. Later in his career Quine certainly did urge a general strategy of turning epistemic questions into matters of empirical psychology. And in his later discussion of natural kinds, he certainly did invoke our psychological propensities to group some things together rather than others. (This last might be thought of as an example of how psychological considerations might determine some choices that are left underdetermined by logical ones alone.) In any case, these concerns with psychology are distinctly later. For now, the idea that there may be a variety of opinions regardless of the evidence does not provide a particularly interesting form of underdetermination. It would have been so obvious to Quine's audience and so obviously irrelevant to their concerns that it is unlikely that Quine would have had this form of underdetermination in mind, at least not this early in his career.

Even if we set the psychological version of underdetermination aside for now, that still leaves at the very least two general possibilities, one, as we said, with simplicity and the other without. To see which of these is more plausibly Quine's we need to reflect on what bearing Quine thought underdetermination has for contemporary science and what role simplicity has in all this. The first thing to notice is that Quine never uses underdetermination to raise a skeptical challenge aimed at current physics. Of course, the claim that there are electrons is fallible, as is the claim from the general theory of relativity that physical space is a non-Euclidian manifold of non-constant curvature. They are underdetermined by the evidence as well. But Quine is perfectly happy to say that they are true. And he seems convinced that we should too. The supposed inadequacy of observational evidence to settle, even temporarily, theoretical disputes has been one of the mainstays of instrumentalism and other forms of anti-realism about theoretical entities. Quine, however, is an avowed scientific realist. So how does one square the idea that there will always be a multiplicity of theories available (underdetermination) with the apparent conviction that at least sometimes one theory is better than the rest? The most plausible answer is that the guaranteed multiplicity of theories is available only in the sense that many theories will fit the observational evidence taken by itself, without invoking considerations such as simplicity. By contrast, the "bestness" of the theory that there are electrons or of Einstein's theory of physical geometry is won only by including simplicity among our considerations. Quine does speak of theoretical science as rounding out and smoothing over the observational evidence. These observational beliefs form only a scattered and gerrymandered part of the simpler whole.

That Quine takes simplicity seriously indeed is reflected as well in his treatment of the claim that there are physical objects. For him this is just a theory, though perhaps coeval with our species, a theory that rounds out and smoothes over the phenomenal evidence (Quine 1948a). Quine takes his physicalism very seriously, and he seems to think that simplicity is a large part of the reason why we should too.

Is there textual evidence from this period in which Quine addresses how to take simplicity? Yes, there is. In 1957 he published "The Scope and Language of Science," in which he said,

> Predictions, once they have been deduced from hypotheses, are subject to the discipline of evidence in turn; but the hypotheses have, at the time of hypothesis, only the considerations of systematic simplicity to recommend them. In so far, simplicity itself – in some sense of that difficult term – counts as a kind of evidence; and scientists have indeed long tended to look upon the simpler of the two hypotheses as not merely the more likable, but the more likely. (Quine 1957, 6)

Now the first full sentence of this was largely cribbed from "Carnap and Logical Truth" written three years earlier. No doubt it was false both times, for at the time of hypothesis there is usually besides simplicity also some observational evidence that can be said to confirm it. The point to note about the quoted passage, however, is that Quine is clearly taking simplicity as on a par with observational evidence and as relevant to the likelihood of a hypothesis. Simplicity and observational evidence are, as one used to say, part of the context of justification. This means that Quine need not, and probably does not, mean by underdetermination that all of our epistemic considerations, including simplicity and conservatism, when taken together, always fail to pick out one hypothesis as better than its rivals. He is more likely to mean instead that the observational evidence, taken by itself, does not entail (via elementary logic) one theory from among its competitors. There is an unavoidable element of conjecture here, of course, given how little Quine says. Still, this interpretation not only has the virtue of plausibility, it is also the way that Quine was read by many of his contemporaries.

There is also a parallel in this with views of Carnap. Carnap saw that physical geometry was underdetermined by the observational evidence alone unless it was supplemented by a method of measurement. But one could not derive such a method from the evidence alone. Simplicity of the overall resultant theory could be invoked to help choose among methods of measurement, though he would have denied that the methodological choice was, properly speaking, uniquely correct. Quine does not draw the theory/framework distinction, and simplicity for him is applied at the same level as is the evidence, but for the issue of underdetermination the result is much the same. When just the observational evidence is taken into consideration, you get underdetermination, especially at the theoretical level; when simplicity is considered as well, you might or might not.

We may summarize, then, the features that we have highlighted about the classic (pre-*Word and Object*) period of Quine's work as follows: Quine rejects outright such intensional notions as analyticity on the grounds that they violate an empiricist criterion of meaningfulness. Correspondingly, he takes reference and ontology as clear and makes them central to his account of language. While Quine's treatment of epistemic matters during this period is both infrequent and sketchy, he plainly has a commitment to a one-tier system, and to underdetermination in the sense lately noted. And consistent with this, he plainly takes simplicity seriously, and on a par with observational evidence and as a way to select one hypothesis as superior to and likelier than its rivals.

10.2 *Word and Object*: The Arguments for Indeterminacy

Let us turn then to what I suggested could be seen as Quine's middle age, that is, to *Word and Object*. Of course, it is a masterpiece that summarizes and rounds out thirty years of thought. But there is a more specific reason for our interest. His first chapter is a substantial methodological/epistemological study, to which we will return. The third through final chapters are that culmination and systematization of his reference-based approach to language. Our focus, however, will be on Chapter II, "Translation and

Meaning." Here he subjected empirical linguistics and especially translation to a very detailed examination. His results were dramatic. The linguists themselves may put forward a translation as uniquely favored by the evidence and argument at their disposal. Even so, Quine says, a great many widely differing manuals of translation could be devised, each of which is completely faithful to all of the behavioral evidence, past, present, and future. Even English can be mapped onto itself in dramatically different ways. This thesis that there are many empirically faultless manuals of translation Quine comes to call the underdetermination of translation. You might think that, even if this much is acceptable, the alien speaker must "have something definite in mind," whether or not the linguist can tell what it is. This Quine denies. He says there is no fact of the matter as to which of the manuals is correct. That the alien has something in mind that would make (at most) one of them correct is just an empty metaphor. This thesis that there is not only underdetermination but no fact of the matter Quine calls the indeterminacy of translation. Needless to say, his indeterminacy thesis was highly provocative and the cause of much controversy. Moreover, it is fair to say that it has never been as widely accepted as his rejection of analyticity. My present purpose is not to appraise indeterminacy or to accept or reject it. Instead, I want to see how it emerged out of Quine's prior thought, what arguments he used to justify it, and what role it played in Quine's subsequent philosophy.

The first thing to note is that Quine's prior rejection of analyticity requires him to say that something must go dramatically wrong with the empirical enterprise of translation, at least if it is thought to discover synonymies by empirical means. Suppose for a moment that nothing went wrong, that is, that for a given alien language linguists could produce a manual of translation that was uniquely favored by the behavioral or other empirical evidence, with or without extra-evidential considerations such as simplicity. The way would then be open to claim that the linguists have empirical criteria for synonymy. But the various terms from the theory of meaning such as 'synonymy', 'meaning', 'entails', and 'analyticity' are, by Quine's own insistence, mutually interdefinable. So if 'synonymy' has sufficient behavioral criteria, then so do all the rest. I don't say that Quine's position on analyticity requires him to claim that there is no fact of the matter about translation, only that something goes wrong in finding a manual of translation uniquely favored by the empirical evidence and argument.

There are in *Word and Object* two different arguments for Quine's conclusions about translation. Both move from underdetermination to indeterminacy via a conception of what a language is, what understanding a language consists in, and how a language can be learned by its native speakers. The two arguments differ in having different routes to underdetermination. Our concern will be primarily with the underdetermination thesis, but we should pause to see what Quine's conception of language is. He makes this explicit in the opening lines to the "Preface":

> Language is a social art. In acquiring it we have to depend entirely on intersubjectively available clues as to what to say and when. Hence there is no justification for collating linguistic meanings, unless in terms of men's dispositions to respond overtly to socially observable stimulations. An effect of recognizing this limitation is that the enterprise of translation is found to be involved in a certain systematic indeterminacy; and this is the main theme of Chapter II. (Quine 1960, ix)

In short, the child has no access to its own language not also afforded to the empirical linguist. If such a linguist cannot tell which manual is correct, then the child cannot either. Since the language can consist only of things that the child can learn about it, there is no fact of the matter. Whatever the merits of Quine's view of language here, he still sees his argument as hinging on underdetermination, and that will be our focus.

Of the two routes to underdetermination one of them can be said to be "from below": there will be a multiplicity of translations because we have a concrete, albeit hypothetical, example (concerning 'gavagai') of two viable translations, and this situation will repeat itself for other languages. The other route, "from above," claims that general considerations about theories show them all to be underdetermined, and a manual of translation is just a special case of a theory. So the translation will be underdetermined as well.

The **argument from below** is better known. Suppose our linguist has progressed to the stage of noticing that every time one points at a rabbit and queries "Gavagai?" the informant is disposed to assent. A great many details have to be slurred over to get to this stage, but for our purposes let us agree not to worry. In such a case, the linguist cannot do better than to translate the one-word observation sentence 'Gavagai' as 'Rabbit' or 'Lo, a rabbit' where those are themselves construed as observation sentences. There is, thus, a natural temptation to translate the term 'gavagai' as 'rabbit' even when it appears in larger contexts, with the remainder of the contexts picking other features of the passing show. Conceivably no violation of the available evidence results from such a translation. But note that every time one points at a rabbit one is also pointing at an undetached part of a rabbit and also at a temporal stage of a rabbit. Even if the translation of 'gavagai' as 'rabbit' is completely successful, there will be other translations of it as 'undetached rabbit part' and as 'temporal stage of a rabbit'. These three alternatives are not only not what anyone would call synonymous in English, they are not coextensive either. The trouble is that differences over the translation of 'gavagai' can be neatly masked by compensating differences in the translations of the remainder of the contexts. One might think that probing questions could be asked of the native so as to determine whether 'rabbit' or 'undetached rabbit part' is meant. Perhaps one might point successively at different parts of the same rabbit while asking "Is this gavagai the same as that gavagai?" That I have had to combine the two languages suggests what is wrong with this approach. If the native is to be queried, the question will have to be in the native's own language. But who is to guarantee in advance that the native expression used in the question in the hope that it meant 'is (numerically) the same as' does not mean 'is attached to' instead? Plainly this difficulty is not restricted to rabbits, their parts, and their stages. Indeed, the further the presumed subject matter is from publicly observable events, the more confident we can be that many translations will each satisfy all of the available data.

How good is this argument from below? In principle, an argument for an existence claim that proceeds by showing us an example is enormously powerful. What could be better? The trouble is that we do not have a well-worked-out example; we do not have two complete manuals of translation even for our hypothetical situation. What we do have is very fragmentary, with the alternatives being carried out only a couple of steps. Quine is quite candid about this. Others such as (Massey 1978) have tried to suggest complete manuals, but have not succeeded to Quine's satisfaction (Quine 1990c, 51). Quine's explanation for the omission is that radical translation is so difficult that if we ever succeeded in producing one manual, no one would be so foolish as to develop a second when there is absolutely no payoff for doing so. Still, the hypothetical fragment we do have is simply no guarantee that it can be carried further. And it is certainly no guarantee that it can be carried so far as to exhaust all of the data that we ever can have. Given these difficulties, the argument from below is perhaps best thought of as a softening-up exercise designed to make us willing to entertain the underdetermination of translation as a real possibility.

The **argument from above** is not fragmentary, but it is abstract. As stated earlier, here the idea is that all theory is underdetermined, and a translation is just a theory. So translation is bound to be underdetermined as well. The general premises of such an argument are laid out in the methodological first chapter of the book. There we are told that the theory of molecules is underdetermined by all the truths that can be said in common-sense terms about ordinary things, and these in turn are underdetermined by the totality of surface irritations (which is Quine's version of the most basic evidence) "even if we include all past, present, and future irritations of all the far-flung surfaces of mankind" (Quine 1960, 22). This claim is made without argument as though it were a long-established theorem. He is less confident about what happens if we throw simplicity and the rest of scientific method into the mix.

But perhaps the whole issue is moot. Quine still emphasizes, at some length, the importance of simplicity, but its character seems to have changed dramatically. We are told: "Yet this supposed quality of simplicity is more easily sensed than described. Perhaps our vaunted sense of simplicity, or of likeliest explanation, is in many cases just a feeling of conviction attached to the blind resultant of the interplay of chain stimulations in their various strengths" (Quine 1960, 19). This suggests that simplicity is vague and subjective, a charge he was to repeat in later years. Worse, simplicity no longer seems to be an independent criterion of belief-worthiness at all; it is just the "blind resultant" of the various forces acting on our beliefs. Quine is not done; later on the same page he says: "Simplicity is not a desideratum on a par with conformity to observation. Observation serves to test hypotheses after adoption; simplicity prompts their adoption for testing" (Quine 1960, 19). In effect, what Quine has done is to demote simplicity from the context of justification to the context of discovery. It might still have a limited role in what hypotheses are ultimately accepted, but it would not be in a position to arbitrate among alternatives.

Though my purpose is not primarily evaluative, it really should be noted that the picture of theory choice that seems to be behind all this is incredibly impoverished. It suggests that for any body of evidence and any two hypotheses, if the two hypotheses are equally simple and each logically consistent with the body of evidence, then they are equally belief-worthy on that evidence. Not only does this fly in the face of nearly all scientific practice, there are simple examples that should give pause to anyone tempted by the view. Let a hundred flips of a coin all of which turned up heads be the body of evidence. Let the two hypotheses be:

1. The probability of getting a head given a flip is 1.0.
2. The probability of getting a head given a flip is 0.0.

where the probabilities in question are both interpreted in the standard limit frequency way. The two hypotheses seem to be equally simple, and they are each logically consistent with the evidence. Yet 1. is plainly the superior hypothesis on that evidence. I don't mean to suggest that this contrived example matches the rich texture of actual practice, but a contrived example is enough for the purpose at hand.

Indeed, whether there will be any underdetermination and if so how much will depend on how underdetermination is defined. It will also depend on what resources one has for sorting hypotheses. The more one adds for preferring one theory over another, the less one gets of underdetermination. Given that Quine's argument for indeterminacy hinges on intension and extension being underdetermined for the alien child, Quine will have to confront psychological forms of underdetermination as well. Finally, the importance of underdetermination will depend on the width of the range of hypotheses left open. This really hurts the argument from above, precisely because it is abstract. Even if underdetermination is taken for granted, such underdetermination provides no abstract guarantee that the alternative manuals of translation will differ in intension and extension as the argument from below seemed to promise. They could differ in some other respects. Quine, of course, gives no argument on any of this, and whether he even can is open to question. So the argument from above does not, by itself, establish anything about reference or meaning.

In effect, Quine is presuming a very strong form of underdetermination. He apparently thinks underdetermination is inevitable when only the empirical evidence is used to select among theories and even if all such evidence, past, present, and future, is used. On the new conception of simplicity, adding that as a consideration would make no difference. Simplicity, on the new account, is relevant to the thinking up of theories, not to the testing of them afterwards, and the two manuals of translation are, by hypothesis, already thought up. So since a manual is just a theory, and all theory is underdetermined, the choice of manuals is underdetermined. To get from underdetermination to indeterminacy, that is to there being no fact of the matter, one need only add Quine's idea that a language consists only of what can be learned of it by initiates who have no more to go on than the linguist does. If this is Quine's argument, and I think it must be, then he could have omitted a great many pages about the intricacies of translation and, of course, undetached rabbit parts. In subsequent presentations of the indeterminacy, he was to do just that. Perhaps the argument from above can be viewed more charitably as reinforcing the fragmentary argument from below by assuring us that the underdetermination by the evidence alone in the first few steps of the "gavagai" case will continue. To extend this charity, perhaps the argument from below can be seen as an attempt, successful or not, to plug the hole just noted in the abstract argument from above, namely by suggesting that within the range of hypotheses left open by underdetermination there are in fact manuals that differ in intension and extension. Whether either of these slender cards can be made to stand by leaning them against each other, we can leave for another occasion.

I said at the outset of our discussion of *Word and Object* that Quine needed something to go wrong with empirical procedures for translation. If something didn't, we would have empirical criteria for synonymy and hence for other intensional terms, and the case against analyticity would have been lost. I think we can assume that Quine would take there being no fact of the matter in translation to be sufficient for his purposes. As so often in life, there was a cost. A minor cost was that the demotion of simplicity threatens the cheerful theoretical realism and physical realism for which simplicity had been a significant support. Of course, Quine goes right on avowing theoretical and physical realism, and he goes right on using simplicity as a main support for this. (Call that Quine's principle of conservatism at work.) But, having demoted simplicity, it is not clear that he is still entitled to do so.

There was another more significant cost. The various empirically adequate manuals of translation could not be counted on to preserve the meaning or intension of the original; it was this about which there was no fact of the matter. Unfortunately, the manuals could not be counted on to preserve the reference or extension of the original either. Quine never tried to disguise the fact that his conclusions included the indeterminacy of reference, that is, that there is no fact of the matter about reference. This really means that the asymmetry between the theory of reference and the theory of meaning that Quine had championed in his classic days is not sustained by his investigation of translation. If the indeterminacy or even underdetermination of meaning shows that the notion of synonymy is not intelligible, then the indeterminacy or even underdetermination of reference should show that the notion of reference is not intelligible either. But don't think that Quine was going to give up on reference just yet; old commitments are hard to change. I shall argue in a bit that the two chapters of *Word and Object* that we have been considering, and especially the tension just remarked on, guided the rest of Quine's career; but the remainder of the book was guided by his previous work. Indeed, the remaining five chapters are the summary and systematization of his reference-based approach to language. The only concession he made to the first two chapters is that reference and the apparatus of quantifiers, identity, and the like are now to be understood parochially, that is, solely within English. That

English can also be mapped onto itself in various ways seems not to have been taken too seriously.

If *Word and Object* is the turning point that I have been promising, that will have to show itself in his later work. Already there is some reorientation on the issue of simplicity, and there is, in the equal indeterminacy of reference and meaning, the basis for still further reorientation. It is to this that we now turn.

10.3   New Directions: Quine's Later Work

Word and Object appeared after thirty years of philosophizing, and there were to be forty more before Quine died in 2000. Indeed, there was new work that came out after that. So it won't be possible to trace all of that work in detail but only to mention a few significant trends and developments. The problem in doing even this is compounded by the fact that especially in this later work there are many distinct currents pulling in different and even discordant directions. Still, we can highlight what is new. And as with the discussion of the Quine of the 1940s and 50s, we can classify these new features as semantic and epistemic.

Quine had wanted a deep asymmetry between notions from the theory of reference, such as 'true' and 'refers', and notions from the theory of meaning, such as 'analyticity' and 'synonymy'. The former were supposed to be legitimate and the later not. The indeterminacy of both reference and meaning, revealed in *Word and Object*, tended to undermine this asymmetry. Quine clung to it nonetheless. But he also defined his own notion of analyticity in *The Roots of Reference* (Quine 1974a, 78–80) and a perfectly adequate notion of synonymy in *Pursuit of Truth* (Quine 1990c, 53). Quine tended to dismiss both of his definitions – prematurely I suspect.

The definition of analyticity (Quine 1974a, 78–80) was couched in terms of language learning, and Quine expressed no reservations about whether there were suitable empirical criteria for it. Instead, his worry was that it was vague and was not extensionally equivalent to Carnap's notion. While it counted elementary logic and truths of essential predication as analytic (as it should), it omitted set theory and parts of mathematics and included claims such as 'there are brown dogs' (as it should not). Even in its present form the definition refutes Quine's previous charge that "Wherever there has been a semblance of a general criterion, to my knowledge, either there has been some drastic failure such as tended to admit all or no sentences as analytic" (Quine 1963a, 403–4) or the definitions were circular or depended on other terms, such as 'meaning', equally lacking behavioral criteria. As for Carnap, he would not have been worried by the vagueness of Quine's notion or by the lack of coextensiveness with his own. He was in the business of giving explications rather than descriptions of English. The whole point of these explications was to precisify and improve on our everyday language.

Quine eventually suggests (in *Pursuit*, 53) that two expressions are synonymous just in case for every body of doctrine large enough to be tested the expressions can be substituted one for the other without disturbing the test conditions. In effect, this is interchangeability *salva confirmatione*. Quine recognizes that this seems right, but he complains that it is vague and does not apply directly between languages. The vagueness Carnap can deal with as before. And even an intralinguistic notion of synonymy is sufficient to define 'analytic', and the latter can be developed into the basis of an account of interlinguistic synonymy as well. One can then get an interlinguistic notion via the account of translation that Carnap gives in *Logical Syntax* (Carnap 1934/1937, 224f.)

Among the semantic features of Quine's later work there is also a continuing concern with reference and ontology. Indeed, the paper that is, so far as I know, his last reaffirms his faith (if that is the word I want) in extensionalism. Reference continues to play a major role, but the discussion is sobered somewhat by the recognition of the indeterminacy of reference and the consequent need to discuss it only from within the provincial confines of the home language. Even the reassuring bromide that 'rabbit' refers to rabbits turns out to be nothing more than an artifact of a homophonic translation, that is, of a mapping of English words onto themselves. Ontology was a somewhat diminished but still active concern, partly perhaps because Quine threw in the towel and accepted sets or classes and partly because, as a result of the indeterminacies, he came to a doctrine of ontological relativity (Quine 1968). Eventually he came to say that what ontology was relative to was a manual of translation (Quine 1990c, 51–2).

In 1988, Quine attended a conference on his work and in his public response to one paper made, to the best of my recollection, the following remark: "I used to think that ontology was central to philosophy, but now I think that it is the epistemology of ontology that plays that role." The remark was never published, and one ought always to be cautious in accepting at face value what Quine says orally but does not publish. Even so, the remark is both right and revealing. After *Word and Object* Quine did have a diminishing concern with ontology, and it was construed as a more provincial and relativized affair. At the same time, his concern with epistemology positively exploded. This is not just *The Web of Belief* (Quine and Ullian 1978) and the whole concern with naturalized epistemology, of which there are many versions and the associated cottage industry. There are studies of simplicity, of underdetermination, and of the supposed multiplicity of theories that could be held, so to speak, till the end of time. This last involved exchanges with Davidson and Føllesdal among others and spanned decades and many changes of opinion (cf. Quine 1990c, 98–100). This greater emphasis on epistemology I am inclined to see as precipitated by *Word and Object*; after all, what was new there hinged on an epistemic claim about underdetermination, and that required a lot of shoring up.

Two aspects of all this deserve special mention. One is that simplicity continued to have the somewhat changed status it was assigned in *Word and Object*. Even as late as *Pursuit of Truth*, simplicity was in a different epistemic category than conformity to observation. Simplicity, which he there calls normative, is relevant only to the thinking up of theories but not to the testing of them afterward. Prediction and conformity to observation he does not see as normative, and they are what testing is all about (Quine 1990c, 20).

The second aspect worth noting is a change that also appears in *Pursuit of Truth*, a change to the basic conception of the epistemological situation as it was made familiar in "Two Dogmas." The change is subtle enough that it seems to escape Quine's own notice there, but even so it profoundly alters the classic Quinean picture. Perhaps he was just checking to see whether we were paying attention. Quine had long since given up the radical holism of "Two Dogmas" according to which only the totality of belief could confront experience. He replaced that totality with suitably large chunks called critical semantic masses, and they were thought of as confronting most directly observation categoricals, that is, generalities compounded of observables (Quine 1990c, 10–12). Still, within any such chunk all our beliefs were tested together. But now he says:

> Over-logisizing, we may picture the accommodation of a failed observation categorical as follows. We have before us some set S of purported truths that was found jointly to imply the false categorical. Implication may be taken here simply as deducibility by the logic of truth functions, quantification, and identity. (We can always provide for more substantial consequences by incorporating appropriate premises explicitly into S.) Now some one or more of the sentences in S are going to have to be rescinded. We exempt some members of S from this threat on determining that the fateful implication still holds without their help. Any purely logical truth is thus exempted, since it adds nothing to what S would logically imply anyway; and sundry irrelevant sentences in S will be exempted as well. (Quine 1990c, 14)

The whole thrust behind "Two Dogmas" was to get logic into the same boat, namely Neurath's, with such mundane beliefs as that there are brick houses on Elm Street. What made the classical Quinean picture exciting is that it promised a one-tier epistemology. Now Quine is back with a two-tier version. There is now a difference with Carnap at most as to where to draw the line between the tiers. And unless Quine has a non-arbitrary and relevant argument as to why it must be drawn at the limit of elementary logic (and not merely that it can be drawn there), he is in some danger of sliding directly into Carnap's view that it is a conventional and pragmatic choice.

Just in case there is any doubt as to whether Quine meant what he said in the passage above about logic not being disconfirmed, he not only repeated it verbatim in the revised edition (1992), he also published the following in 1995:

> Treating observation categoricals as empirical checkpoints of scientific theory, in section I above, I evidently gave logic a role separate from the rest of science. If a set of theoretical sentences is tested by testing an observational categorical that is implied by the set, then surely the logic of the implication is no part of the tested set. But how much logic? Where we demarcate logic, for purposes of the implication is evidently an artifact of our analysis and not intrinsic to the scientific method under analysis. (Quine 1995b, 352)

Plainly, the interpretation I gave of the first passage is something that Quine himself intended.

These semantic and epistemic changes are in the aggregate quite dramatic. Before *Word and Object*, Quine pushed for a deep asymmetry between reference and meaning in which analyticity and synonymy were unintelligible while reference and ontology were central. And he placed little emphasis on epistemology. And within that Quine's commitments were to a one-tier system, to underdetermination, and to a conception of simplicity that was central to theory choice. After *Word and Object*, the earlier reference-meaning asymmetry was undermined by the indeterminacies, by giving definitions of analyticity and synonymy that met his earlier criticisms, and by relativizing ontology. Moreover, epistemology was emphasized, a two-tier system was offered in which logic played a special role, and simplicity was reconceived and demoted.

But what is new often stands side by side with older elements to which they are quite antithetical. I suspect that this is in part because his view changed faster and more profoundly than his self-conception. Though his view came increasingly to resemble Carnap's, he continued to see himself in contraposition to Carnap, or more precisely, in contraposition to a certain (mis)conception of Carnap. He continued to see himself basically as still the author of "Two Dogmas." And, of course, he was. There are real continuities in Quine's work, early, middle, and late. In saying that *Word and Object* was a turning point in Quine's philosophy, I don't mean to suggest that he made a U-turn. It was no mid-life crisis. It was more of a mid-course correction. As noted, Quine was fifty-two when he published that book, ripe middle age. That is hardly the time for a revolution, even if Quine's principle of conservatism would have allowed it. It was more nearly the time for just what *Word and Object* turned out to be: in the last five chapters a brilliant rounding out of and summation of Quine's career thus far, and in the first two chapters the spur for innovations to come. It was, in short, a turning point.

The book itself is now well over fifty years old, and perhaps our understanding of it has likewise reached middle age. Certainly it is time for reflection and review. In my remarks here I have attempted to provide some of that here, and also a mid-course correction, if you will. But with apologies to Doris Day, our philosophic understanding need not outgrow its middle age, its period of reflection and self-correction. With luck, that will have a beginning and a middle, but not an end.

PART IV

Carnap and Quine on Ontology and Metaphysics

CHAPTER II

Carnap and Quine on Ontology and Categories
Roberta Ballarin

This essay joins the debate around Quine's understanding of Carnap's "Empiricism, Semantics, and Ontology" (*ESO*). In *ESO* Carnap puts forward his views on ontology and kinds.[1] The essay's tone is initially defensive. Carnap upholds his own theoretical use of abstract entities, primarily propositions and properties in semantics, against the attacks of fellow empiricists of a nominalistic bent, such as Quine. But the essay ends with a counterattack. Carnap famously concludes that it is not his own use of abstract entities in semantics that must be regarded as a problematic form of old-fashioned metaphysical Platonism. It is rather the attitude of his attackers, Quine *in primis*, that is based on an old-fashioned ontological attitude. For Carnap, the real metaphysical sin lies not so much in the scientific endorsement of abstract entities, but in the attribution of theoretical meaningfulness to the old ontological Realism–Nominalism dispute, no matter which side one supports.[2] Carnap develops the notion of a linguistic framework to bolster this counterattack. Quine responds to *ESO* in "On Carnap's Views on Ontology" (*CVO*). In this essay I consider the question whether Quine's reading of *ESO* as concerned with the category/subclass distinction is just a misunderstanding of Carnap's views. I argue that it is not, thus supporting the core insight of Quine's interpretation of *ESO*, which puts special emphasis on category questions of existence.[3]

---

[1] I defend the claim that *ESO* is concerned with kinds in the rest of the paper. For now, just consider Carnap's own heading to Section III of *ESO*: "What Does Acceptance of a *Kind* of Entities Mean?" (213; emphasis added).
[2] See Alspector-Kelly (2001) on this aspect of the Carnap–Quine debate, and on the evolution of Quine's ontological views, starting with an intuitive endorsement of nominalism and ending with "the question whether quantification over abstract entities can be eliminated from scientific discourse without harm to the scientific endeavor" (2001, 97). Carnap, argues Alspector-Kelly, regards Quine's early nominalistic intuitions as a problematic synthetic a priori stand on ontology.
[3] At the two extremes, Bird (1995) thinks that Quine grossly misunderstands Carnap. Ebbs (2017, 2019) fully supports and deepens Quine's reading of *ESO*.

In the recent scholarly debate, most commentators object to Quine's reading of Carnap, in particular to the following two theses of Quine's: (i) that Carnap's external questions of existence are all category questions; and (ii) that answers to internal category questions of existence are trivial and analytic. Recently, Ebbs (2017, 2019) has defended Quine on both points. This essay supports Ebbs' conclusion on the first point. But the epistemic considerations I employ in support of Quine's first point undermine Quine's second proposal. So, I will argue that the answer to internal category questions of existence cannot be trivial. Even if we concede to Quine and Ebbs that the acceptance of a linguistic framework comes with a trivial endorsement of the new kind of entities, I will argue that this trivial endorsement cannot be the answer to an *internal* category question, in a strong sense of "internal." Key to my take on these two points is an understanding of frameworks as fundamentally branches of science, each with its own distinct methodology.

It is clear that both Carnap and Quine struggled to figure out what exactly their disagreement consisted in, and Quine wrote *CVO* with this task in mind. Yet all their ontological disagreements notwithstanding, Carnap and Quine appear to agree on one major assumption, namely, that *to be is to be the value of a variable*. Indeed, Carnap concedes that "Quine was the first to recognize the introduction of variables as indicating the acceptance of entities" (*ESO*, 214, n. 3). This essay argues that though Carnap endorses the letter of Quine's famous motto, it is exactly on the interpretation of this motto that the crux of their disagreement lies. Carnap endorses Quine's words, but interprets them in a radically new way, which differs in essential respects from Quine's original idea. This essay argues that we need to be aware of how differently Carnap and Quine interpret the idea that to be is to be the value of a variable in order to reach a deeper understanding of their disagreements on ontology, as well as a better appreciation of Quine's reading of *ESO*.

11.1 Linguistic Frameworks for New Kinds of Entities

In *ESO* Carnap claims that in order to understand the nature of kind-specific ontological questions, such as "Are there properties, classes, numbers, propositions?" (206) one must distinguish between two types of questions regarding the existence of such entities: "If someone wishes to speak in his language about a new kind of entities, he has to introduce a system of new ways of speaking, subject to new rules; we shall call this procedure the construction of a linguistic *framework* for the new entities in question" (206; emphasis in original). Once a framework for a new kind of entities is set up, Carnap distinguishes between existence questions asked from within the framework and existence questions "concerning the existence or reality *of the system of entities as a whole*" (206; emphasis in original), which he deems external.

Carnap is very clear that each linguistic framework is set up for a specific kind of entities. Frameworks are not devised to address all sorts of questions of existence, but questions of existence for "a new kind of entities" (213). This seems to exclude at the very least questions regarding the existence of anything whatsoever as opposed to nothing; questions of the existence of one particular thing, such as my left hand, irrespective of its kind; as well as questions of the existence of a plurality of things of a disparate variety of kinds. Once a framework for the new kind of entities is in place, one will be able to address questions regarding the existence of particular entities of that new kind by simply applying the rules of the framework. In contrast, external questions addressing the existence of the new kind of entities in their entirety (the system of entities as a whole) are regarded as "devoid of cognitive content" (212), that is to say, as "pseudo-questions" (213).

In *ESO* Carnap provides many examples of frameworks for distinct kinds of entities: the world of things, for "the simplest kind of entities dealt with in the everyday language" (206); the system of numbers, for natural numbers; the system of propositions; the system of thing properties; the system of integers and rational numbers; the system of real numbers; and the spatio-temporal coordinate system for physics. Carnap is explicit that different linguistic frameworks are associated with different kinds of entities: "The acceptance of a new kind of entities is represented in the language by the introduction of a framework of new forms of expressions to be used according to a new set of rules" (213). Indeed, Carnap maintains that the two essential elements of a framework are "first, the introduction of a general term, a predicate of higher level, for the new kind of entities, permitting us to say of any particular entity that it belongs to this kind (e.g., 'Red is a *property*', 'Five is a *number*'). Second, the introduction of variables of the new type. The new entities are values of theses variables" (213). Inessential elements of a framework are, first, the proper names for things of the new kind, for example, numerals such as "five" for numbers and names such as "Caro" for ordinary (material) world things; and, second, the predicates that need not apply to all the entities of the new kind, such as "prime" and "odd" for numbers, "dog" and "brown" for things. So, the essential elements of a framework appear to be those that characterize its entities as *of a certain kind* (natural numbers, material things, propositions, properties, etc.), as well as their corresponding types of variables, such as "n," "x," "p," or "f," respectively.[4]

Questions internal to a framework are addressed within the framework already set up. They presuppose the framework and look for answers within it, that is, according to its system of rules. Carnap mentions that a framework comes with a new set of rules associated with its terms, but we shouldn't think of these as just syntactic or semantic rules for the use of terms. A framework must include rules of an epistemic sort too, meant to give proper procedures or methods to answer questions about the existence of specific (subclasses of) entities within the new kind, for example, dogs and unicorns among things, prime numbers greater that 100 among numbers, colors among properties, and so on. Carnap is explicit on this need for epistemic procedures when describing the world-of-things and the system-of-numbers frameworks. Of the first he says that internal questions such as "Is there a white piece of paper on my desk?":

> are to be answered by empirical investigations. Results of observations are evaluated according to certain rules as confirming or disconfirming evidence for possible answers. (ESO, 207)
> To accept the thing world means nothing more than to accept a certain form of language, in other words, to accept rules for forming statements and for testing, accepting, or rejecting them. (ESO, 208)

Similarly, the system of numbers is characterized as containing numerals (names of numbers), the general term "number," expressions for properties of numbers, their relations and functions, as well as rules to form sentences, and finally numerical variables and the existential and universal quantifiers "with the customary deductive rules" (208). In this case too, the framework includes not only terms and rules to build sentences but also a proof procedure. In the case of the system of numbers, the proof procedure is a deductive formal apparatus, not an empirical methodology for confirmation or disconfirmation, but in both cases the framework contains more than just a language to speak of the new entities. Frameworks include an epistemology too, understood as a methodology for proving or disproving, confirming or disconfirming that there are specific entities of

---

[4] Given these explicit remarks of Carnap's, it is somewhat puzzling that scholars should have rejected Quine's suggestion that Carnap's frameworks separate first and foremost category, namely, *kind* questions that pertain, so to speak, to the construction of the framework, from subclass questions that are answered internally (*CVO*, 207). Bird (1995) aims to support Carnap while agreeing with Quine that the category/subclass dichotomy is untenable. It becomes then essential for him to deny that Carnap's external/internal distinction is somehow dependent on an (allegedly) untenable one.

the new kind, for example, keys in my pocket and tigers in Alabama, or multiples of 17 between 80 and 100.

Is the epistemological component essential or not to a framework? It is clear that it must be, as, if it weren't, a framework would be sorely inadequate to fulfill the task for which it was designed, that is, to address the question of the existence of a new sort of entities in a nonmysterious, scientific, anti-metaphysical way. It is exactly this epistemic or methodological component – the framework's epistemology – that bestows cognitive content on internal questions. In contrast, the metaphysical question whether numbers have "a certain metaphysical characteristic called reality" cannot be answered because it is not supplied with "a clear cognitive interpretation" (209). This must be the case because metaphysicians have not devised a shared scientific epistemology for tackling the question whether a number satisfies the predicate "having reality." Mathematicians instead follow a clear method to prove whether a number is prime; and regular folks know, though they cannot necessarily spell it out, how to figure out whether there is a white piece of paper on their desk. It seems then that the epistemic/methodological component is essential to a framework. Scientists have not only clear languages, but relatively clear epistemologies too, epistemologies that they tend to agree upon. Metaphysics' main problem consists exactly in the fact that its practitioners cannot even agree on their discipline's methodology. If frameworks lacked a relatively well-defined epistemology, they wouldn't be much better off than metaphysics.[5]

Notice, moreover, that according to Carnap, it is not only the case that the framework for the world of things provides us with empirical methods to answer existence questions for ordinary material things. The framework also supplies us with "an empirical, scientific, nonmetaphysical concept" of reality:

> The concept of reality occurring in these internal questions is an empirical, scientific, non-metaphysical concept. To recognize something as a real thing or event means to succeed in incorporating it into the system of things at a particular space-time position so that it fits together with the other things recognized as real, *according to the rules of the framework*. (*ESO*, 207; emphasis added)

---

[5] The Carnap–Quine meta-ontological dispute has re-emerged in contemporary metaphysics. According to Eklund (2009 and 2013), the two main contemporary understandings of a framework are a framework as a language vs. a framework as a worldview or perspective. Eklund is unsatisfied – rightly so, it seems to me – with these two interpretations. This paper argues that frameworks are neither just a language nor a perspective on reality. Frameworks are more like branches of science, each with its own specific methodology (though they don't necessarily exactly correspond to actual branches of science); consequently, Carnap's take on ontology is not so much a plea for shallowness, but an attempt to make it scientific. Metaphysicians and historians (some more than others) are surely aware of the epistemic component of frameworks, but I think more emphasis needs to be put on it.

Metaphysicians endlessly debate not just whether there really exists a world of things, but also what the reality of the material world consists in. The world-of-things framework ends both fruitless debates, on the existence of material things and on the nature of material existence, by incorporating a clear answer to the question of material reality: "to recognize something as a real thing or event means to succeed in incorporating it into the system of things at a particular space-time position so that it fits together with the other things recognized as real, according to the rules of the framework." This passage strongly suggests that the metaphysical, unclear concept of the material reality of ordinary things is to be replaced by the scientific epistemological concept implicitly determined by the rules of the framework. To be a material thing is not to possess some mysterious form of material reality or existence. It is instead to be incorporated in a spacetime position within the whole system of material entities, and this is something that can be established empirically.[6] Similarly, to be a (natural) number is not to enjoy "a kind of ideal reality, different from the material reality of the thing world" (209). It is rather to be the kind of entity that can be incorporated into the system of natural numbers and whose existence can be established by arithmetical calculation. Naturally, the relevant frameworks do not *state* that being part of the spacetime structure of things is what the material reality of a thing consists in, or that being designated by a sentence is what the reality of a proposition consists in. Rather, such answers are, so to speak, implicit in the epistemology employed by the respective frameworks. Yet if – as I have argued – a framework requires an epistemology, why doesn't Carnap list it among its *essential* features? When singling out the essential features of frameworks, Carnap mentions only (i) the general term for the new kind of entities and (ii) their distinct type of variables (in contrast to names and special predicates). How can this be? I will return to this question towards the end of Section 11.2.

11.2 Kinds, Variables, and Rules

Having clarified what a framework consists in, how it is related to the existence of a new kind of entities, and why it must incorporate a scientific epistemology, let us turn to Quine's analysis of *ESO*. Quine introduces the *category/subclass* distinction as more basic – in Carnap's own system – than the *external/internal* opposition:

---

[6] One might worry on the circularity of this criterion, but such a concern would only apply if the criterion were meant, as it obviously is not, to be a theoretical definition of what "being material" consists in.

It begins to appear, then, that Carnap's dichotomy of questions of existence is a dichotomy between questions of the form "Are there so-and-so's?" where the so-and-so's purport to exhaust the range of a particular style of bound variables, and questions of the form "Are there so-and-so's?" where the so-and-so's do not purport to exhaust the range of a particular style of bound variables. Let me call the former questions *category* questions, and the latter ones *subclass* questions. I need this new terminology because Carnap's terms 'external' and 'internal' draw a somewhat different distinction which is derivative from the distinction between category questions and subclass questions. The external questions are the category questions conceived as propounded before the adoption of a given language; and they are, Carnap holds, properly to be construed as questions of the desirability of a given language form. The internal questions comprise the subclass questions and, in addition, the category questions when these are construed as treated within an adopted language as questions having trivially analytic or contradictory answers. (*CVO*, 207)

Given Carnap's explicit remarks that frameworks are for kinds of entities, Quine's proposal that Carnap's most basic dichotomy is between category (that is, kind) questions and subclass questions seems spot on.[7] Quine picks the term "category" for questions concerned with the existence of (the entire system of) all the entities of the new kind. Such questions, claims Quine, may be raised in different ways. They may be conceived as external questions, propounded prior to and independently of the linguistic framework for the entities in question. To ask ontological questions in such a metaphysical way is anathema to Carnap, for the reason we already discussed, that externally raised questions are unaccompanied by a clear methodology. For Carnap, the only questions one may legitimately raise externally are practical questions on the desirability of the linguistic framework.[8] For Carnap, the scientific philosopher shouldn't raise category existence questions at all. He should rather just build a framework that will incorporate a methodology that, as we have seen, somehow implicitly addresses them (as Carnap himself did for semantics) or, when appropriate, accept the answer implicit in an already existing framework (as, for example, in the case of arithmetic). We have also seen that the epistemology of the framework will include the proper procedures to answer questions regarding the existence of some entities of the new kind, that

---

[7] Ebbs (2019) upholds Quine's reading of *ESO* and argues that Carnap relies on the notion of universal concepts introduced in *The Logical Syntax of Language* and further developed in *Meaning and Necessity* §10, where universal concepts are linked to variables.
[8] We may wonder why the epistemological standards for practical, as opposed to theoretical, questions are lower. Why can practical questions be addressed with no clear methodology in place?

is, subclasses of such entities, such as dogs or unicorns, even prime numbers greater than 2, color properties, and so on. So, subclass questions are indeed internal questions.

What is controversial in Quine's analysis are the claims (i) that external questions just are category questions asked from without a framework, thus ruling out external subclass question; (ii) that category questions can also be asked internally; and (iii) that when a category question is raised internally its answer is either "trivially analytic or contradictory."[9] The rest of this section is devoted to claim (i).

The problem with claim (i) arises because even if Carnap elucidates external questions as concerning the existence of the entire system of entities, and thus as category ones, Carnap's key worry with external questions does not seem to concern their generality but rather their lack of a guiding epistemology; and it seems clear that one might raise in a similarly unguided, metaphysical, unscientific way any question of existence whatsoever. So, against Quine and the letter of Carnap's text – after all Carnap does say that external questions concern "the existence or reality of the total system of the new entities" (214) – one may still advocate that external questions need not concern the entire new kind. However, Ebbs (2019, 11–14) argues convincingly that the epistemic standpoint of someone who worries, externally, only about the real existence of a subclass of, for example, material entities (let's say, tables), while remaining satisfied with internal answers for what concerns the existence of other subclasses of material things (let's say, chairs and lamps), is unstable. Such a standpoint implies the simultaneous acceptance and rejection of the framework's internal methodology: acceptance, insofar as chairs and lamps are concerned, yet rejection for tables.

Clearly, the instability of this position depends on the fact that chairs and tables are entities of the same kind, that is, ordinary material things, and as such included in one and the same framework. The position of a philosopher who worries, externally so to speak, about the real existence of numbers, but not of material things, is not equally unstable. But what does it mean to be entities of one and the same kind? What are kinds? We can only expect that Carnap's answer to this philosophical question won't be fruitlessly theoretical. To elucidate Carnap's take on it, we need to return to the essential elements of frameworks. Given that frameworks are built to answer questions about the existence of kinds of entities, their essential

---

[9] Doubts are typically voiced against (i) and (iii). I have added (ii) for reasons that will become clear later. See Bird (1995) against (i) and (iii).

features will reveal Carnap's take on kinds. I am applying to Carnap his own recommendation, but at a more general level. We have seen that for Carnap we shouldn't philosophically inquire on the nature of the material reality of things. We should instead figure it out based on the epistemology built within the specific framework for material things – and similarly for the ideal reality of numbers. In an analogous way, it seems that rather than enquiring on the nature of kinds and kind distinctions, one should instead figure out what kinds are by attending to the essential features of linguistic frameworks as such. We don't have a framework for kinds, so we cannot attend to a specific framework for them. But all frameworks are built to handle a specific kind of entities, so attending to the features that are essential to all frameworks will reveal what it means to recognize something as a kind.

As seen, Carnap states that the essential parts of a framework are its fundamental general term ("thing," "number," "property of things," etc.) and the corresponding variables. One question looms large: both the general terms and the variables are quite arbitrary. The fundamental distinction between ordinary material things and numbers or properties surely cannot depend on, or be reduced to, such arbitrary typographical accidents. It's hard to believe that all there is to being a thing versus a number is to be in the domain of "$(\forall x)$" versus "$(\forall n)$." It is equally hard to believe that all there is to being a kind/category of entities as opposed to a subclass thereof is to exhaust the domain of a special type of variables. Quine's objection to Carnap along these lines is well known. Yet it is important to properly recognize the complexity of Quine's main objection to Carnap, which does not simply consist in pointing out the arbitrariness of typography. Quine claims that typographical distinctions are trivial, and criticizes Carnap exactly because he seems to deny this triviality:

> Carnap does not just have this trivial distinction in mind. He is thinking of languages which contain fundamentally segregated styles of variables … and he is thinking of styles of variables which are sealed off from one another so utterly that it is commonly ungrammatical to use a variable of one style where a variable of another style would be grammatical. A language which exploits this sort of basic compartmentalization of variables is that of Russell's theory of logical types. (*CVO*, 208)

So, the main problem is not so much that the compartmentalization of variables is a typographical accident, but that Carnap doesn't seem to recognize this; and the objection to Carnap seems to be that if not merely typographical, then the distinction can only be ontological. Quine in fact proceeds to object to Russell's theory of types. Some of Quine's objections are no doubt based on the idea that the choice of segregated variables is, for him, a typographical accident, but Quine is aware that ultimately what needs to be abandoned is "Russell's notion of a hierarchical universe of entities disposed into logical types" (209). Indeed, if Russell were right on the logical stratification of the universe, variable segregation wouldn't be just an accident of typography; it would instead correspond to a real distinction. This, I take, is Quine's main attack on Carnap: the rigid separation between frameworks, encapsulated in their distinct variable types, is wedded to something like a logically stratified ontology. The implicit suggestion is that, like Russell, Carnap is committed to a universe of ontologically segregated kinds of entities, that is, entities that occupy distinct layers of reality.[10]

If my interpretation is correct, we see that once again Quine accuses Carnap of a metaphysical sin: Platonism to start with, and now an ontology of segregated kinds, each one enjoying its own sort of existence or reality. We may say that Quine has remained (intentionally?) deaf to the principal point of *ESO*, which attempted to silence all charges of this sort. Carnap's reply to the accusation of Platonism was that if this sort of complaint stems from a nominalistic ontological insight, it won't carry much weight. Better arguments need to be devised (221).[11] Insofar as Quine's objection to the segregation of variables is of a similar sort, that is, insofar as the rejection of distinct logical or ontological types is based on an ontic intuition, it too won't carry much weight. Carnap can thus reiterate his counterattack: it is Quine who is still in the grip of an old ontological dispute. Moreover, Carnap would, of course, deny being committed to a problematic ontology of kinds, just as he denies being committed to a problematic form of Platonism. In fact, Carnap's multitude of separate frameworks, each with its sealed variables, is exactly designed to replace these objectionable metaphysical distinctions between different modes of reality or existence (recall Carnap's remarks on the material being of things and the ideal being of numbers).

Similarly to the Platonism–Nominalism dispute, in the case of ontic categories too, Quine and Carnap lean toward the same side, at least in the sense that they both find metaphysical Platonism and a stratified ontology,

---

[10] Thus, I disagree with Thomasson's assessment of Quine's discussion of Russell's theory of types as an irrelevant detour: "But although it occupies the vast majority of this influential article, the discussion about styles of variables, and category versus subclass questions, is really a technical sideshow distracting from the real metaontological issues" (2016, 129).

[11] Alspector-Kelly (2001) points out that Quine's later nonintuitive stand on ontic decisions, e.g., in *Word and Object*'s last chapter, agrees with Carnap's. See note 2.

understood as encompassing different forms of existence, objectionable; but they react differently. Quine leans toward nominalism[12] and frankly denies different kinds of existence, being, or reality for different kinds of entities:

> There are philosophers who stoutly maintain that "true" said of logical or mathematical laws and "true" said of weather predictions or suspects' confessions are two usages of an ambiguous term "true." There are philosophers who stoutly maintain that "exists" said of numbers, classes, and the like and "exists" said of material objects are two usages of an ambiguous term "exists." What mainly baffles me is the stoutness of their maintenance. What can they possibly count as evidence? Why not view "true" as unambiguous but very general, and recognize the difference between true logical laws and true confessions as a difference merely between logical laws and confessions? And correspondingly for existence? (1960, 131).[13]

Carnap instead introduces a linguistic counterpart to the metaphysical doctrine. The frameworks are designed to replace nonsensical distinctions between distinct ontic categories with scientifically respectable segregated languages. The frameworks do not characterize different kinds or categories of entities based on ontic intuitions about the sort of entities that populate their separate domains: material things for "x" and "y," numbers for "m" and "n." Which kind of entities populate the range of the variables of a framework must be determined by an element internal to the framework itself, and this can only be the epistemology of the framework. If there is no prior ontology presupposed, and if the distinction is not merely typographical, it can only be the rules of a framework that determine its domain, that is, the kind of entities that populate its universe. That is why to be a material thing is to be the kind of entity whose existence is confirmed or disconfirmed by the empirical procedures that either place it or fail to place it in the spatio-temporal system of things. In general, to be an entity of a certain kind is not to enjoy a characteristic form of reality, but to be subject to a certain set of epistemic rules. This, I surmise, must be Carnap's ultimate reply to Quine. Once again, we confirm that at the core of a framework lies its epistemology. I conjecture that the reason why Carnap does not directly mention the epistemology of a framework as essential is that the epistemology is what determines the domain of the framework. But distinct domains are represented by distinct variables and the concomitant universal term. Distinct types of variables are thus essential to a framework, not as mere signs, but insofar as they signal the framework's specific methodology.

---

[12] Though he does not think nominalism is achievable; see Quine (1960, §48).
[13] Similar sentiments permeate Quine's "On What There Is."

It is then very clear that the position of someone who raises an external question concerning the existence of a subclass of entities is unstable. If it makes sense to raise a special worry about a subclass of entities, it must be the case that such entities belong to a special sub-category worthy of the creation of its own more specific framework. That is why it is simply silly to raise special doubts about the existence of tables but not chairs, while it seems perfectly appropriate to raise special doubts about the existence of artifacts or of living organisms in opposition to other less complex material things. These metaphysical queries, however, would be addressed by Carnap with the construction of corresponding frameworks nested within the general world-of-things one. If there is a special science for living organisms, and there is, with its proper vocabulary and methodology, it is its job to ponder on their existence. Similarly, special concerns on the existence of artifacts need to be addressed not by means of ontic intuitions but with the construction of a clear linguistic framework.

Similarly, if we build a unified framework for things and numbers, we may, if we want, adopt one type of variables and one common universal term for all the entities in the combined framework, but what we cannot do is pretend that the same kind of epistemic considerations would be adequate to establish the existence of all the entities in the domain of this mixed framework. Such a mixed language may be useful for ordinary conversation, but the unified variables will obscure the fundamental epistemological distinctions between the science of numbers and the "science" of ordinary things. These epistemological distinctions are what the category distinction between numbers and things amounts to. In the unified language, the terms "number" and "material thing" would be special subclass predicates and we would regard it as false, rather than nonsensical, to say of a number that it is blue and of a chair that it is a multiple of 2. But what we could not give up is the two distinct methodologies for proving or disproving the existence of the entities in the distinct subcategories. A legitimate concern about the existence of numbers would appear superficially as a concern about the existence of a subclass of entities, but this alleged subclass is indeed a masked category of its own. The first of Quine's claims, that for Carnap external questions are category questions, seems vindicated.[14]

---

[14] See Ebbs' (2019) illuminating interpretation of both *ESO* and *CVO*. Ebbs seems to be putting more emphasis on frameworks as logical rather than epistemological systems, thus concluding that Carnap's segregation of variables is less crucial than I take it to be.

11.3 Internal Category Questions

We can now turn to Quine's additional claims, that category questions can also be internal and that when they are so propounded their answers are either trivially analytic or contradictory. Intuitively, this very last claim is controversial insofar as it seems clear that the existence of empirical entities such as physical objects is an empirical matter. Nobody seems to doubt Quine's assertion that both category and subclass questions can be internal, and Carnap does indeed characterize some category questions as internal. The disagreement concerns Quine's assertion that when asked as internal questions, category questions have trivially analytic answers. The following passage from *ESO* appears to support Quine's reading:

> To begin with, there is the internal question which, together with the affirmative answer, can be formulated in the new terms, say, by "There are numbers" or, more explicitly, "There is an n such that n is a number." This statement follows from the analytic statement "five is a number" and is therefore itself analytic. Moreover, it is rather trivial ... because it does not say more than that the new system is not empty; but this is immediately seen from the rule which states that words like "five" are substitutable for the new variables. (*ESO*, 209)

Most scholars, however, disagree with Quine's suggestion that all internal category questions are trivially analytic or contradictory, and claim that internal category questions are empirical in, so to speak, empirical frameworks.[15] As Ebbs (2019, §4) points out, this has become the "New Standard Reading" of *ESO*. Against the New Standard Reading, Ebbs (2019, §7) argues that an argument analogous to the one given by Carnap for the analyticity of "there are numbers" can be constructed for the world-of-things framework, thus concluding that Carnap is committed to the analyticity and triviality also of the internal category assertion "there are physical things" (internal, of course, to its corresponding framework). Ebbs strengthens his argument by the additional consideration that the detour through names in the above argument is irrelevant. This is important as Carnap regards names as inessential to a framework. And, we must add, even if names were essential, the dispute between Ebbs and the New Standard Reading would focus back on the analyticity of "Caro is a (material) thing," with Ebbs granting it and the New Standard Reading denying it.[16]

---

[15] See Bird (1995, 55); Yablo (1998, 236); Soames (2009, 429); Alspector-Kelly (2001, 106); and Thomasson (2016, 127).

[16] Ebbs (2019, 19–20) quotes a passage from *The Logical Syntax of Language* (293) where speaking of universal terms Carnap points out that "Caro is a thing" is analytic. But the question remains open whether Carnap would still say the same at the time of *ESO*, and more crucially whether he should say so.

To bypass the issue of names, Ebbs argues that it is open to Carnap to accept a meaning postulate stating that there are numbers or that there are ordinary physical things as part of the corresponding frameworks. Indeed, in the system-of-numbers framework where variables range over numbers only, the postulate "there is an n such that n is a number" can be expressed as "there is an n such that $n=n$." Similarly, in the world-of-things framework, "there is an x such that x is a thing" can be expressed by "there is an x such that $x=x$." Ebbs also argues convincingly that this strategy of trivializing internal category statements of existence was already part of Carnap's plan of eschewing metaphysical questions and replacing them with the construction of linguistic systems at the time of *The Logical Syntax of Language*. Positive answers to such questions are built into the linguistic systems and reinterpreted as quasi-syntactical or (later) semantical, rather than as genuine assertions in the material mode. Thus, Ebbs argues, the assumption that the domain of a linguistic framework is not empty is part and parcel of the framework itself. In support of the plausibility of regarding statements of existence as analytic (according to Carnap's understanding of analyticity), Ebbs reminds us of Quine's own preference for classical first-order logic with its assumption of a nonempty domain, based on pragmatic considerations of technical simplicity and a deflated interpretation of logical truth somehow endorsable by Carnap too (see Quine 1953).[17] Ebbs seems to be suggesting that the theoremhood of "$\exists x\, x=x$" carries over from first-order logic to Carnap's frameworks. If instead we adopt one unified domain (as Quine would prefer), the assumption of existence for distinct categories of entities will be supplied by kind-specific meaning postulates stating "there are numbers" or "there are physical objects," and so on.[18]

I find Ebbs' arguments insightful and to some degree compelling. However, they are *prima facie* in conflict with other passages of *ESO*. For example, when considering a sentence such as "there are colors" in the framework of the properties of things, Carnap claims that this "sentence is an internal assertion. It is of an empirical, factual nature" (212). Now, if this is an empirical assertion, and if we can derive from it that there are properties of things (as colors are), then the category statement that there are properties, taken as internal, seems itself empirical. Of course, this consideration

---

[17] Quine denies that logical truths are true in virtue of meaning; Carnap accepts that they are but in his own revolutionary understanding of this notion, which – claims Ebbs – leaves space for "$\exists x\, x=x$" being true in virtue of meaning.

[18] This is a brief and somewhat rough reconstruction of Ebbs' complex defence of Quine's claim, but I hope not to have misrepresented its key points.

does not contradict the possibility that "there are properties" be ultimately analytic, but there seems to be some tension between two different interpretations of the statement. Similarly, if we establish that Caro is a dog, based on empirical investigations, and then conclude on its basis that Caro is a physical thing (rendered as "Caro = Caro"), and so that there are physical things, what we establish seems more substantial than the analytically true "Caro is a thing" (and "there are things") of *The Logical Syntax of Language*, which was deemed analytic but also devoid of content ("its L-content is null and it is analytic," *The Logical Syntax of Language*, 293). Perhaps it is devoid of content to say that something exists, but is it similarly devoid of content to say that something is a property or a physical object?

Consider also the following passage from *ESO*:

> The critics of the use of abstract entities in semantics overlook *the fundamental difference between the acceptance of a system of entities and an internal assertion*, e.g., an assertion that there are elephants or electrons or prime numbers greater than a million. *Whoever makes an internal assertion is certainly obliged to justify it by providing evidence, empirical evidence in the case of electrons, logical proof in the case of prime numbers. The demand for a theoretical justification, correct in the case of internal assertions, is sometimes wrongly applied to the acceptance of a system of entities.* ... Some nominalists regard the acceptance of a system of entities as a kind of superstition or myth, populating the world with fictitious or at least dubious entities, analogous to the belief in centaurs or demons. This shows again the confusion mentioned, because a superstition or myth is a false (or dubious) internal statement. (218; emphasis added)

This passage seems to be in clear conflict with the one that argued for the triviality of the *internal* statement "there are numbers." Indeed, it opposes internal assertions of existence to the (trivial) acceptance of the whole system of entities. Previously, we were told that the internal statement "there are numbers" is trivial and analytic insofar as "it does not say more than that the new system is not empty" (a statement that strongly support Ebbs' Quinean interpretation of *ESO*), and also that the triviality of internal category assertions of existence is the reason why those who doubt such a statement cannot mean it internally:

> So philosophers who ask this question as a serious philosophical question do not mean it internally. And, indeed, if we were to ask them: "Do you mean the question as to whether the framework of numbers, *if* we were to accept it, would be found to be empty or not?," they would probably reply: "Not at all; we mean a question *prior* to the acceptance of the new framework." (*ESO*, 209)

In contrast, however, Carnap is now claiming that *internal statements must be justified* "by providing evidence, empirical evidence in the case of electrons, logical proof in the case of prime numbers." He also lists only subclass examples of internal assertions, explicitly opposing them to the acceptance of the whole system: "The demand for a theoretical justification, correct in the case of internal assertions, is sometimes wrongly applied to the acceptance of a system of entities." That internal statements must be theoretically justified confirms our suggestion that its epistemology is an essential, indeed *the essential*, element of a framework, typographically represented by the segregation of variables and the concomitant introduction of a universal term for the new entities. Justification is exactly what separates internal scientific statements from their external philosophical counterparts.[19] But if justification is necessary to internal statements, then the *internal* statements that there are material things, or properties thereof, cannot be trivial in the strong sense of standing in need of no justification, nor can they be analytic, as the forms of justification of these frameworks are empirical. Qua internal statements, they too must be empirically justified. Similarly, and crucially, even the statement that there are numbers, if it is derived from the statement that there are prime numbers greater than 200 (for example), has been proved. As such, given the character of its justification, it counts as analytic, but not as trivial in the philosophically relevant sense.[20]

It seems then that *ESO* contains two distinct and contrasting sorts of considerations about internal category questions of existence. On the one hand, there is the idea, emphasized by Ebbs, that a linguistic framework incorporates the assumption that entities of the new kind exist. This line of thought pushes in the direction of regarding internal category questions as trivially true. On the other hand, insofar as linguistic frameworks are no mere logico-linguistic systems, but contain an essential epistemology, all internal questions, even those concerning the existence of entities of the most general category or kind, will have to be internally justified in accordance with the framework's own epistemology. They will then, against Quine and Ebbs, have to be empirical (for things) or analytic (for numbers), and in no case trivial.

---

[19] Though, as seen, the only philosophically interesting external questions are category questions.

[20] Carnap's proof of "there are numbers" from "five is a number" is a different and special case, given that the referentiality of "five" – but not the existence of primes – is built into the structure of the linguistic framework. So understood, the statement is trivial and built into the framework to start with, but is not internally justified. The question of how to draw the distinction between analytic statements that pertain to the construction of the framework of the system of numbers and those that are derived internally is, unsurprisingly, a very difficult one.

How can we reconcile these two strands of thought? It seems that there are two distinct notions of *internal* in place:[21] (i) internal in the weak sense of simply nonexternal; and (ii) internal as internally justified. The subclass statements of existence, and the category ones derived from them, are internal in the strong sense of depending for their justification on substantial epistemological considerations. Instead, a trivial internal category question of existence depends for its answer on the simple construction of the framework, what we might perhaps regard as its pure logico-linguistic component, before the internal process of justification is, so to speak, activated. As seen, Carnap himself seems to be using the term ambiguously. Quine and Ebbs seem to be using "internal" in the first, weak sense, while the New Standard Interpretation is using "internal" in the second, strong sense.[22] We may say that the New Standard Reading is right for existence questions under a strong interpretation of "internal," but in a weak sense of "category," while Quine and Ebbs are right for existence questions conceived in a weak sense of "internal" but a strong sense of "category" – concerning the original positing of the whole system of entities, not its derivation based on subclass internal statements.

However, according to Carnap, category questions, in the strong sense of "concerning the reality of the whole system of entities," are either external or deceivingly theoretical:

> [W]e take the position that the introduction of the new way of speaking *does not need any theoretical justification because it does not imply any assertion of reality*. We may still speak (and have done so) of "the acceptance of the new entities" since this form of speech is customary; but one must keep in mind that this phrase does not mean for us anything more than acceptance of the new linguistic forms. Above all, it must not be interpreted as referring to an assumption, belief, or assertion of "the reality of the entities." There is no such assertion. An alleged statement of the reality of the system of entities is a pseudo-statement without cognitive content. ... *The acceptance cannot be judged to be either true or false because it is not an assertion*. (ESO, 214; emphasis added)

These remarks of Carnap's seem to address exactly Quine's suggestion that category questions of existence are either problematically philosophical or trivial, namely, in no need of scientific theoretical justification. Category existence questions are trivial in the sense of being implicitly answered by the sheer acceptance of the linguistic framework, and are thus

---

[21] Ebbs (2017, 37) proposes so much, but does not develop this point.
[22] This is not to say that their disagreement is merely verbal.

not internally theoretically justified; but they are also trivial in the sense that the acceptance of the new entities amounts to nothing more than the acceptance of the new linguistic forms. However, this pragmatic, and as such theoretically unjustified, acceptance of the new entities ought not to be regarded as a theoretical statement. Thus, it is at least misleading to suggest that category questions of existence are trivially analytic, as this suggests that they are trivially analytically *true*.[23]

In contrast, genuinely internal assertions of existence are theoretical statements. As such they stand, as we have seen, in need of justification, as no scientific statement can be regarded as immune from justification. If the segregation of variables is taken seriously, then to be the value of a variable of the framework of things is not simply to be an existent in a neutral logical sense. It is instead to be a physical object in the scientific sense embodied in the methods of confirmation and disconfirmation of the framework. This cannot be a merely logical trait. The same, of course, is true for abstract entities, including numbers, unless, of course, logicism is correct.[24]

11.4 To Be Is to Be the Value of a Variable

In closing *CVO*, Quine declares himself hopeful that he might persuade Carnap to abandon the category/subclass distinction, and make do with just the analytic/synthetic one:

> No more that the distinction between *analytic* and *synthetic* is needed in support of Carnap's doctrine that the statements commonly thought as ontological, viz., statements such as 'There are physical objects', 'There are classes', 'There are numbers' are analytic or contradictory given the language. No more that the distinction between analytic and synthetic is needed in support of his doctrine that the statements commonly thought of as ontological are proper matters of contention only in the form of

[23] Ebbs seems to downplay the nontheoretical status of these pseudo-assertions: "In what sense is 'acceptance of the thing world' not a theoretical assumption, in Carnap's view? The answer, I suggest, is that such acceptance is an analytic consequence of adopting what Carnap calls the thing language" (2017, 40).

[24] If the epistemology embedded in a framework, its methods of confirmation or proof, is kept separate from its language, then we may distinguish the logical from the substantially epistemological part of the framework. In this case, the trivially analytic "assertions" of existence will be merely logical. But if the epistemology of the framework is sewn into the semantic rules of its terms, *including the category term*, then it is hard to concede that internal category questions can have trivially analytic answers. These considerations on the distinction between the logical and the frankly epistemological part of a framework intersect in a complex way with the difficult question whether the trivial assertions of existence are to be conceived as genuine assertions of actual existence or the mere positing of the transcendental possibility of the existence of entities of that kind.

linguistic proposals. The contrast which he wants between those ontological statements and empirical existence statement such as 'There are black swans' is clinched by the distinction of analytic and synthetic. *True, there is in these terms no contrast between analytic statements of an ontological kind and other analytic statements of existence such as 'There are prime numbers above a hundred'; but I don't see why he should care about this.* (CVO, 210; emphasis added)

I don't read Quine as saying that the category/subclass distinction can be reduced to the analytic/synthetic one. On the contrary, Quine seems to be proposing that Carnap should abandon the category/subclass distinction, and let go of the additional concomitant distinction between ordinary analytic statements (of existence) and trivial analytic statements concerning the existence of ontological categories.[25] We see here that Quine reiterates his reading of *ESO* as principally concerned with statements of existence *of an ontological kind*, that is, statements of existence of fundamentally distinct categories of entities. Quine does not endorse the analytic/synthetic distinction, as he rushes to remind us, but the suggestion to Carnap is to let go of an additional questionable dichotomy.[26] Quine is as unsympathetic as ever to Carnap's philosophical project of reforming dubious metaphysical notions, such as necessity, meaning, and now category distinctions, by explicating them in linguistic terms. As for nominalism, in the case of the denial of ontological categories too, we can imagine that Carnap's response to Quine will be that more than sheer intuition is required. To persuade Carnap, Quine needs to show that science is indeed unified to the extent that no clear separation of methodologies is justifiable – quite a tall order to fill!

In closing the essay, I would like to reiterate that *ESO*'s main task seems to be to explicate in linguistic/epistemological, anti-metaphysical terms not just the bare logical notion of existence, but also the notion of belonging to distinct ontological categories. Carnap endorses Quine's claim that to be is to be the value of a variable. But what do they mean by this? Quine is always explicit that his concern is not with what exists but with what exists according to a theory:

---

[25] This distinction appears already in *The Logical Syntax of Language* where Carnap contrasts "standard" analytic statements and trivial analytic *genus* statements (293).
[26] According to Bird, Quine objects to Carnap's external/internal distinction first by reducing it to the category/subclass one and then to the analytic/synthetic one, so for Quine "Carnap's distinction requires, or can be made in terms of, the contrast between analytic and synthetic truths" (1995, 47). Ebbs seems to agree with Bird, but endorses this reducibility. In contrast, I don't read Quine as claiming that the analytic/synthetic distinction suffices for capturing the category/subclass distinction. Rather, Quine is urging Carnap to let go of additional ontological distinctions that the analytic/synthetic dichotomy won't suffice to capture.

Now how are we to adjudicate among rival ontologies? Certainly the answer is not provided by the semantical formula "To be is to be the value of a variable"; this formula serves rather, conversely, in testing the conformity of a given remark or doctrine to a prior ontological standard. We look to bound variables in connection with ontology not in order to know what there is, but in order to know what a given remark or doctrine, ours or someone else's, says there is; and this much is quite properly a problem involving language. But what there is is another question. (1948a, 34–35)

Not for Carnap! For Carnap "to be real in the scientific sense means to be an element of the system" (*ESO*, 207), and there is no prior ontological standard to match. There is no question of what there is over and above what there is according to a theory, a linguistic system, or framework, and there is no question of being over and above the question of being the value of a variable. We may therefore say that for Carnap to be *just is* to be the value of a variable.[27] Not only so, if I am right that Carnap's project in *ESO* is to explicate ontological categories too by similar linguistic-epistemological means, we may also conclude that for Carnap *to be a number is to be the value of a numerical variable, to be a thing is to be the value of a thing variable,* and *to be a proposition is to be the value of a propositional variable*; where the values of a type of variables are determined by the scientific methodology spelled out by the epistemic rules of the corresponding framework. Unlike Quine, Carnap supplants questions of existence as possession of reality with questions of existence as scientific recognition. As a consequence, metaphysical distinctions between the forms of reality enjoyed by distinct ontological kinds of entities are replaced by the epistemic distinctions between the separate branches of science. Where Quine prefers to eliminate ontological categories, Carnap's proposal in *ESO* is to reinterpret them in a scientifically viable way.

---

[27] Although this way of making the point is too metaphysical, I'd like to say too Quinean, to be to Carnap's liking.

CHAPTER 12

Carnap and Quine on the Status of Ontology
The Role of the Principle of Tolerance

Peter Hylton

As is well known, Carnap and Quine disagree over the status of ontology. Quine holds that the question 'What is there?' is a genuine, substantive question, and that answering it is part of the philosopher's remit; Carnap, by contrast, holds that what appear to be ontological questions are pseudo-questions, or else questions of a quite different kind stated in a misleading form. My concern in this essay is with the basis of this disagreement. Both authors hold that it rests on their disagreement over the analytic–synthetic distinction. Quine, however, comes to accept a version of that distinction, and his doing so does not in any way modify his view of the status of ontology. This gives us reason to ask: what other ideas are crucial for Carnap's position on the status of ontology, and are disputed by Quine? The answer, I argue, is the Principle of Tolerance. The analytic–synthetic distinction by itself, without Tolerance, does not provide a basis for a Carnapian view of the status of ontology. The fact that Quine comes to accept a version of the distinction is heuristically useful, as a way of enabling us to see this fact; but it is not essential to the argument that Tolerance, as well as the analytic–synthetic distinction, is necessary to sustain the Carnapian view.

12.1 Carnap and Quine: The Disagreement over the Status of Ontology

Carnap and Quine disagree over the status of ontology. This is not a disagreement over substantive ontological questions as to what exists. It is, rather, a disagreement over whether ontology is a genuine subject at all. Carnap's view is that it is *not* a genuine subject. He claims that supposed ontological assertions are pseudo-statements: sequences of words which look like real assertions worthy of scientific investigation, but which are in fact nonsensical, making no real claim about the world. In the 1934 essay 'On the Character of Philosophic Problems' he takes the dispute over the existence and nature of numbers as an example and speaks of 'metaphysical pseudo-problems concerning the nature of numbers, whether the numbers are real or ideal, whether they are extra- or intramental [sic] and the like' (Carnap 1934c, 14f).

The same attitude, and the same idea that supposed ontological questions are pseudo-questions, persists after Carnap's adoption of semantic methods, as well as syntactic methods (and, indeed, after his move to the United States, and to writing in English). In the 1950 essay 'Empiricism, Semantics, and Ontology' he says that we are justified in taking 'the philosophical question concerning the existence or reality of numbers' as 'a pseudo-question'.[1]

Quine, by contrast, treats ontological questions as genuine questions. In the early (1948b) essay 'On What There Is' he famously says: 'A curious thing about the ontological problem is its simplicity. It can be posed in three Anglo-Saxon monosyllables: "What is there?" It can be answered, moreover, in a single word – "Everything" – and everyone will accept this answer as true' (1). The answer does not, of course, settle the question of what exists. As Quine immediately goes on to say: 'There remains room for disagreement about cases' (*loc. cit.*). But he takes it for granted that there is a real question. There is no suggestion here that we have a pseudo-problem; indeed, the idea of a *disagreement* precisely suggests the opposite, that there is a genuine question to which different philosophers give different answers. And in the rest of the essay Quine goes on to argue for substantive answers to the question. He argues, for example, against the acceptance of entities which subsist but do not exist (such as were accepted by, for example, Russell in the *Principles of Mathematics*), and against the acceptance of entities which are possible but not actual. Along the way he makes various methodological remarks which make it clear that he is discussing what he takes to be a genuine question. Thus, he says: 'We commit ourselves to an ontology containing numbers when we say there are prime numbers larger than a million' (8) and, more generally, 'To be assumed as an entity is ... to be reckoned as the value of a variable' (13). By the time of *Word and Object*, he is arguing for quite definite ontological theses: that only sets and physical objects (in his sense) exist.

So Quine certainly accepts that ontological questions are real questions, and sets out to answer them. His acceptance of the indeterminacy of reference, otherwise known as ontological relativity, makes no difference here; I will briefly elaborate on this point, although it takes us a little way from the main line of argument.

---

[1] Carnap (1950/1956), 209; quoted in context, 242, below.

Quine first sets out the doctrine of the indeterminacy of reference in *Word and Object*. According to that doctrine, there is more than one acceptable translation of the referring terms of a language; the various translations may be equally correct even though they are not merely stylistic variants of one another. This doctrine does not, however, undermine the idea that ontological questions are genuine questions, with answers which are correct or incorrect: it merely means that the answers are not unique. When I use the term 'my favourite copy of *Word and Object*', I can be translated either as talking about a certain set of space-time points (those which that book occupies), or as talking about the space-time complement of that set, that is, the set of all other space-time points. The indeterminacy of reference tells us that either translation may be acceptable, given appropriate adjustments elsewhere; indefinitely many other translations may also be acceptable. But it does not affect the idea that our theory of the world, suitably regimented, *has* an ontology. The indeterminacy of reference does nothing to undermine the idea that ontological questions are real; it provides no reason to accept the Carnapian idea that ontological questions make no sense. So the contrast between Quine's view and Carnap's remains.

Further considerations suggest that the indeterminacy of reference does not make a crucial difference to ontology. First, it is not just the translation of one term in a sentence that is affected; to make a given translation of that term acceptable, the translations of the other words in that sentence must also be modified in suitable fashion. This fact sharply constrains the variety of acceptable translations of the referring terms. More than one translation may be acceptable, but not every translation will do. Second, and perhaps more strikingly, the indeterminacy of reference does not change the ontology that Quine attributes to our theory as a whole. The general constraints that he puts on ontology are unaffected. In particular, the slogan 'No entity without identity', which embodies Quine's great objection to propositions and properties and other creatures of darkness, is unchanged. He holds that physical objects and sets comprise what there is; his reasons for holding this are not affected by the indeterminacy of reference. The term 'my favourite copy of *Word and Object*', as I use it, can be translated as referring to one set of space-time points or to another; but all the space-time points, and all the sets thereof, are in our ontology either way.

In the last two paragraphs, I have been arguing that Quine's acceptance of the indeterminacy of reference does not take him any closer to Carnap's view of the status of ontology. Quine continues to think of ontological questions as genuine questions, and to put forward his preferred answers to them: the disagreement between him and Carnap over the status of ontology is unaffected. In spite of Quine's doctrine of the indeterminacy of reference, there is thus a clear and sharp difference between him and Carnap over the status of ontological questions. Carnap dismisses them as pseudo-questions; Quine takes them to be real questions, deserving an answer.

12.2 The Basis of the Disagreement

What does this disagreement over the status of ontology rest on? Quine himself, in an essay published in 1951, says quite explicitly that it rests on the differing views that the two philosophers have over the analytic–synthetic distinction:

> No more than the distinction between *analytic* and *synthetic* is needed in support of Carnap's doctrine that the statements commonly thought of as ontological are proper matters of contention only in the form of linguistic proposals. The contrast which he wants between those ontological statements and empirical existence statements such as 'There are black swans' is clinched by the distinction of analytic and synthetic. (1951a, 210; emphasis in the original)

Carnap agrees with Quine that their disagreement over the analytic–synthetic distinction underlies and explains their disagreement about the status of ontology. In 'Empiricism, Semantics, and Ontology', he emphasises the distinction between 'the acceptance of a linguistic framework' and 'a metaphysical doctrine concerning the reality of the entities in question' (214). In a note appended to this passage, he says: 'Quine does not acknowledge the distinction which I emphasise above, because according to his general conception there are no sharp boundary lines between logical and factual truth, between questions of meaning and questions of fact' (215, n. 5).

So Carnap and Quine both hold that their disagreement about the status of ontology is based on their disagreement about analyticity. But Quine's views about analyticity change considerably over time, or at least appear to do so; he seems to be far less hostile to the idea in 1974, and later, than he was in 1951. In *Roots of Reference*, he puts forward a more or less behaviouristically grounded definition of analyticity. For each person, there are certain sentences which he or she came to accept in the course of learning to understand them. As an example, Quine gives a sketch of how we might come to understand sentences of the universal categorical form, and says: 'If we learned to use and understand "A dog is an animal" in the way I described, then we learned at the same time to assent to it, or account it as true' (1974a, 79). For each person, there will be some sentences of which this holds: that he or she came to accept the sentence in the course of first learning to understand it. Moreover, Quine claims, there will be some sentences that have this characteristic for every user of the language. And those, he says, are the sentences that should be counted as analytic: 'Here then we may at last have a line on the concept of analyticity: a sentence is analytic if *everybody* learns that it is true by learning its words' (1974a, 79, emphasis in original).

In the 1991 essay 'Two Dogmas in Retrospect', he makes the point even more clearly. He endorses the earlier definition, and also suggests that we may regard the set of analytic truths as closed under analytic transformations: 'truths deducible from analytic ones by analytic steps would count as analytic in turn' (1991, 396). By this criterion, he suggests, all of elementary logic, that is, first-order logic with identity, may be analytic: 'All logical truths in my narrow sense – that is, the logic of truth functions, quantification, and identity – would then perhaps qualify as analytic, in view of Gödel's completeness proof' (1991, 396).[2]

Quine's views about analyticity thus seem to change. What are the implications of this apparent change for his views about ontology? Does his friendlier view of analyticity go along with an acceptance of Carnap's view that what look like ontological claims are pseudo-statements? Or with any concessions at all to that view? It seems clear that the answer to these questions is No. After 1974 Quine continues to treat ontology as a genuine subject, just as he does earlier.

As we saw at the start of this section, Quine says that his rejection of Carnap's view of ontology rest on his opposition to the analytic–synthetic distinction. As his views develop, however, he backs away from that opposition, at least to some extent. But this apparent shift in his views on analyticity is not accompanied by any change in his view of the status of ontology. The obvious question is: why not?

To answer this question, it will be useful to go over Carnap's rejection of ontological questions, to see exactly what it rests on, and work back from there. Approaching the disagreement over ontology in this way will show that acceptance of the analytic–synthetic distinction is necessary for Carnap's way of rejecting ontology. It will also show that the Principle of Tolerance, or something to the same effect, plays a crucial role in supporting Carnap's views about the status of ontology.

---

[2] For a more detailed account of Quine's post-1974 acceptance of a notion of analyticity, see Hylton (2021).

12.3 Carnap's Reasons for Holding that Ontological Questions are Pseudo-Questions

So let us rehearse Carnap's reasons for holding that ontological questions are pseudo-questions. From the early 1930s on, Carnap holds that there is a multiplicity of possible languages, which are not merely variants of one another, but which differ in expressive power and, in some cases, even in underlying logic. Given that there are various languages, it is presumably possible to switch from one to another, to give up speaking one language and to adopt a different one. So we can distinguish two kinds of questions. On the one hand, there are questions in which we take for granted the language we are speaking and ask whether – given that we are speaking that language – such-and-such a sentence is true. Carnap sometimes speaks of these as *internal* questions, since they arise within a language. On the other hand, there are what Carnap calls *external* questions, of which the opposite is true: in such a case, the questioner does not take the language for granted but, to the contrary, means to cast it into doubt.[3]

The internal–external distinction is not the same as the analytic–synthetic distinction. A question about an analytic sentence might be asked as an internal question. Suppose we are speaking a language in which arithmetic is analytic, and someone raises the question whether there is a prime number between 10,000 and 10,010. The answer is analytic, but the question would naturally be interpreted as internal, one that does not seek to cast doubt on the language we are speaking but simply asks whether, given that we are speaking that language, it is correct to say that there are prime numbers within the given range. A question such as 'Are there numbers?' (or perhaps 'Are there any numbers *really*?'), by contrast, is not naturally interpreted as an internal question because, construed in that way, it is completely trivial. Someone who asks that question is unlikely to be asking the trivial question; it is far more likely that they intend to ask a question of a different kind, an external question, one which does not take the language for granted. A similar contrast can be drawn for synthetic sentences. Someone who asks whether there are wolves in Michigan is presumably asking an internal question, not meaning to cast doubt on the language that we are speaking. Someone who asks whether there are physical objects, by contrast, is naturally interpreted not as asking a trivial question – one

---

[3] For a more detailed account of Carnap's views on this topic, see the work cited in the previous footnote, Section 12.3.

that can be answered by kicking a stone or taking one's hands out of one's pockets – but rather as asking a question of a different kind.[4]

The difference between an internal question and an external question is thus a matter of what kind of answer would satisfy the questioner. The same form of words may be used to raise either kind of question. If the question is an internal one, then the questioner will be satisfied by an answer which presupposes the rules of the language. This holds whether the answer to the question is an analytic sentence or a synthetic sentence. In cases of the former kind, the answer to the question will be established by the rules of the language alone; if we are speaking a language in which arithmetic is analytic, those rules are all that is needed to show that 10,007 is prime. In cases of the latter kind, when a synthetic sentence is in question, we will need both the rules of the language and empirical information. As Carnap says: 'Results of observations are evaluated according to certain rules as confirming or disconfirming evidence for possible answers' (1950/1956, 207). Evaluating observations in that way will show us that there are indeed wolves in Michigan.

If the questioner is not satisfied by an answer which presupposes the rules of the language, then she is not asking an internal question. This is the case with ontological questions. The philosopher who asks whether there are numbers is unlikely to be satisfied by being told that it is analytic that 'five is a number', and that the statement 'there are numbers' follows from 'five is a number' (compare 1950/1956, 209). Likewise for synthetic sentences: the philosopher who raises the question as to the existence of physical objects is unlikely to be satisfied by being shown a pair of hands and being assured that hands are physical objects. In Carnap's terms, philosophers' ontological questions are external questions.

We should not think of Carnap's external questions as being legitimate questions which are simply of a different kind from internal questions. Having explained internal questions, he does not proceed to a parallel explanation of external questions. He says, rather, 'An external question is of a problematic character which is in need of examination' (1950/1956, 206). He then considers what he calls 'the thing language', the language in which we talk about ordinary things and events; he gives examples of ordinary questions which we raise within the framework of that language. He contrasts those internal

---

[4] I assume here that the statement that there are physical objects would count as synthetic for Carnap. Gary Ebbs has argued that 'general existence statements', as Carnap views them in 'Empiricism, Semantics, and Ontology', are by Carnap's lights analytic. See his 'Carnap on Ontology', ch. 3 of his (2017). I take no stand on this issue (which presumably depends on the details of the language under consideration); my assumption is purely for the purposes of exposition.

questions with 'the external question of the reality of the thing world itself' (1950/1956, 207). This supposed question, he says, 'is raised ... only by philosophers', and gives rise to disagreements that are never settled (*loc. cit.*). The question, he says, 'cannot be solved because it is framed in the wrong way. To be real in the scientific sense means to be an element of the system; hence this concept cannot be meaningfully applied to the system itself' (*loc. cit.*).

The point here is that raising a question presupposes a framework. The framework gives rules for answering questions (most obviously: for carrying out proofs, in the case of a question which has an analytic answer; for evaluating evidence, in the case of a synthetic answer). In raising a (supposed) external question, the philosopher is abrogating those rules, making the question she is apparently raising senseless. Carnap suggests a charitable interpretation:

> Those who raise the question of the reality of the thing world itself have perhaps in mind not a theoretical question as their formulation seems to suggest, but rather a practical question, a matter of a practical decision concerning the structure of our language. We have to make the choice whether or not to accept and use the forms of expression in the framework in question. (1950/1956, 207)

Carnap makes essentially the same point about what he calls 'the system of numbers', meaning the language and the logic in which we do arithmetic, and speak and reason more generally about numbers. He considers the question 'Are there numbers?' No one who raises that question means it as an internal question, that is, one which takes for granted the system of numbers; taken in that way it is too trivial for anyone to be interested in it. Carnap envisages those who raise the question as saying that they mean '"a question *prior* to the acceptance of the new framework" ... the question whether or not numbers have a certain metaphysical characteristic called reality' (1950/1956, 209). He responds:

> Unfortunately, these philosophers have ... not succeeded in giving to the external question and to the possible answers any cognitive content. Unless and until they supply a clear cognitive interpretation, we are justified in our suspicion that their question is a pseudo-question, that is, one disguised in the form of a theoretical question while in fact it is non-theoretical; in the present case it is the practical question whether or not to incorporate into the language the new linguistic forms which constitute the framework of numbers. (1950/1956, 209)

The rest of section 2 of Carnap (1950/1956) makes similar points about other systems of entities: propositions, rational numbers, real numbers, and space-time points; section 3 generalises the point. All that is needed for the acceptance of entities of a new kind, according to Carnap, is 'the introduction of new forms of expression to be used according to a new set of rules' (1950/1956, 213). In particular, he says, we need a new general term ('number', 'property', etc.) and a new type of variable.[5] These resources will enable us to formulate internal questions about the new entities, and to evaluate possible answers to them. These internal questions are quite distinct from 'external questions, i.e. philosophical questions concerning the existence or reality of the total system of the new entities' (1950/1956, 214). He dismisses (supposed) questions of this latter sort:

> 'An alleged statement of the reality of the system of entities is a pseudo-statement without cognitive content. To be sure, we have to face at this point an important question; but it is a practical, not a theoretical question; it is the question whether or not to accept the new linguistic forms' (1950/1956, 214).

Ontological questions are thus pseudo-questions, in Carnap's view, because philosophers misunderstand them. The philosopher who asks whether there are physical objects does not think that she is asking a question with a wholly trivial answer. But neither does she think that she is merely asking about which language to speak. Such a philosopher, or at least the kind of metaphysician that Carnap seeks to counter, takes herself, rather, to be asking a genuine theoretical question, neither trivial nor simply about language. Of course, we actually speak a language in which we talk about physical objects. But are there in fact any such objects out there, independent of us and of our language? Are there physical objects *really*? That is the metaphysician's question. And it is exactly the sort of question that Carnap wants to dismiss as a pseudo-question, as lacking significance. Any question, he insists, is either internal or external. If our philosopher's question is taken as internal, then it is trivial; if it is taken as external, then it is a practical question about language choice, not a theoretical question about the supposedly language-independent issue of the existence of physical objects.

12.4 Where Does Quine Disagree?

Assume, for the moment, that this more or less represents the movement of thought which leads Carnap to hold that ontological questions are pseudo-questions. We can then ask: where does Quine disagree? He does not disagree with the idea that various non-equivalent languages are

---

[5] Quine argues that the requirement for a new style of variable is misplaced, even by Carnap's own lights. See Quine (1951a), 207–10.

possible. He was certainly aware of the fact that more than one underlying logic is possible, and into the 1950s he continues to speak as if a sense-datum language were a genuine possibility.

Nor does Quine disagree with the idea that what objects one accepts is in large part a matter of what language one has adopted. In 'On What There Is' he says

> One's ontology is basic to the conceptual scheme by which he interprets all experiences ... Judged within some particular conceptual scheme – and how else is judgment possible? – an ontological statement goes without saying, standing in need of no separate justification at all ...
>
> Judged in another conceptual scheme, an ontological statement ... may, with equal immediacy and triviality, be adjudged false. (1948b, 10)

A little later in the essay, he makes the point more explicitly: "Disagreement in ontology involves basic disagreement in conceptual schemes ... It is no wonder, then, that ontological controversy should tend into controversy over language" (1948b, 16). Quine's view here seems to be very much akin to Carnap's on the same point.

As we saw in Section 12.2, both Carnap and Quine attribute their disagreement over the status of ontology to their disagreement over the analytic–synthetic distinction. We emphasised in Section 12.2 that the internal–external distinction cannot simply be identified with the analytic–synthetic distinction. Nevertheless, the two distinctions are importantly related. In particular, the internal–external distinction depends upon the analytic–synthetic distinction. If there is no clear distinction between questions of meaning and questions of fact, then there is no clear distinction between change of language and change of theory within a language.

Adapting an example of Putnam's will help to make the point. Suppose that one scientist accepts that the kinetic energy of an object at a given moment is half the product of the mass of the object and the square of its velocity at that moment ($e = \frac{1}{2} mv^2$), and perhaps calls this statement a 'definition'; suppose also that another scientist rejects that claim (see Putnam 1962, 42). Is this a disagreement over a factual claim within a single language? Or is it best understood as a proposal to change languages? The issue turns on what language we should best construe the first scientist as speaking, and whether '$e = \frac{1}{2} mv^2$' is analytic in that language. (Carnap would presumably hold that there are some languages in which it is, and some in which it is not.) It is crucial for Carnap's view that for any given language there is a definite answer to the question whether the sentence is analytic: if there is no clear answer to that question, then, equally, there is no clear answer as to whether rejecting the sentence involves a change of language or merely a change of theory within the same language.

This point holds, of course, for all sentences. If we do not, in each case, have a clear answer to the question of the analyticity of the sentence (relative to a given language), then, equally, we have no clear answer to the question whether a change of verdict on the sentence would be a change of language. In that case, the distinction between change of language and change of theory within a language collapses, and with it Carnap's internal–external distinction. And without this latter distinction, Carnap's argument that ontological questions are pseudo-questions also collapses. So the analytic–synthetic distinction is crucial to the disagreement between Carnap and Quine over the status of ontology.[6]

Since Quine comes to accept that some truths are analytic, however, he also comes to accept that in some cases a change of theory *does* involve a change of meaning. In 'Two Dogmas in Retrospect' he writes:

> If the logical truths are analytic – hence true by the meanings of words – then what are we to say of revisions, such as the imagined case of the law of the excluded middle? Do we thereby change our theory or just change the subject? This has been a recurrent challenge, and my answer is that in elementary logic a change of theory *is* a change of meaning. Repudiation of the law of the excluded middle would be a change of meaning, and no less a change of theory for that. (1991, 396; emphasis in the original)

Quine backs away from his opposition to the analytic–synthetic distinction; he thus comes to accept that we can, at least in some cases, make a clear distinction between change of language and change of theory within a language. So he gives up on the most obvious basis for his opposition to Carnap's use of the internal–external distinction. Yet there is no sign that he becomes any more willing to agree with Carnap that ontological questions are pseudo-questions. Hence our question remains: what basis does the post-1974 Quine have for continuing to reject Carnap's view of the status of ontology? To answer this question, we need to consider what *other* presuppositions of Carnap's argument Quine might object to.

12.5 Other Points of Disagreement

In summing up Carnap's argument, I put it this way: if an ontological question is taken as internal then it is trivial. But if it is taken as external, then it is a question about language, not about the supposedly language-independent issue of the existence of physical objects. That is not wrong,

---

[6] I have argued this fairly straightforward point at some length because it has been denied. See, in particular, Eklund (2013) 229–49.

exactly, but it is too quick. Let us grant Carnap the distinction between internal questions and external questions, and grant also that ontological questions are external, that is, that they concern the choice of language rather than the choice of theory within a language. It does not immediately follow that such questions are not genuine substantive questions, with right and wrong answers. To get that result, we need to add the claim that the choice of language itself is not a genuine substantive question. If one holds the opposite view – if one thinks, for example, that some languages reflect the true nature of reality and others do not – then one will have no difficulty in taking ontological questions to be *both* questions about language *and*, at the same time, genuine substantive questions. Whether there are numbers (as fundamental entities) may be settled by our choice of language. But if the question of language choice is itself a substantive one, with a right answer, then the ontological question will also have a right answer. If the *correct* language – the one whose structure reflects the nature of the world, we are supposing – contains number terms as primitive, then the ontological claim that there are numbers is justified.

To support Carnap's conclusion about the status of ontology, then, we need more than the distinction between internal questions and external questions. We also need to deny the idea that there is such a thing as the *correct* language. And this Carnap does, in what has come to be called the Principle of Tolerance. According to this principle, the notion of correctness is applicable to the choice of a theory within a given language (i.e., to internal questions) but not to questions of language choice. Questions of the latter kind are not a matter of correctness, not a matter of right and wrong at all. One formulation of the principle, from the *Logical Syntax of Language*, reads as follows:

> '*In logic there are no morals.* Everyone is at liberty to build up his own logic, i.e. his own form of language, as he wishes. All that is required of him is that, if he wishes to discuss it, he must state his methods clearly, and give syntactical rules instead of philosophical arguments' (1934/1937, 52).

Given this principle, and the internal–external distinction, Carnap's argument about the status of ontology is hard to resist. Which language we speak is a matter of free choice, not a matter of right or wrong to be settled by philosophical argument. Since ontological claims are trivial, given the language, no genuine issue arises about such claims. If I say that numbers are fundamental entities, distinct from sets, then a philosopher who sets out to contradict me is either denying something which is trivially true, given the language we are both speaking, or is speaking a different language. In the latter case, we may have a simple misunderstanding: we think we are speaking the same language, when in fact we are not. If there is to be a dispute here, rather than a misunderstanding, it presumably arises because my opponent thinks that I am wrong to speak the language that I am speaking, and should speak her language instead. But that position contravenes the Principle of Tolerance. According to that principle, I am free to choose whatever language I wish, and if I choose to speak one in which the existence of numbers is a triviality, that is certainly my right.

12.6 A Russellian Alternative to the Principle of Tolerance

So the Principle of Tolerance is crucial to Carnap's view of the status of ontology. Given this fact, one might wonder about alternatives to that principle. The alternative position which Carnap seems to have in mind is Russell's, from before 1918. According to that view, all our knowledge is based on a direct and immediate cognitive relation between the mind and certain entities; entities given in sensory experience are the most obvious example, but Russell also includes abstract entities. He calls this relation *acquaintance*.

Acquaintance is fundamental to Russell's view of how we understand language. The language in which we think, the language which is revealed by the process of analysis, is one in which every term refers to an entity with which the speaker of the language is acquainted. If this does not hold for a language, then the speaker cannot understand it: linguistic understanding of a term is fundamentally a matter of being acquainted with the entity for which the term stands.[7] As Russell puts it: 'The fundamental principle in the analysis of propositions containing descriptions is this: *Every proposition which we can understand must be composed wholly of constituents with which we are acquainted*' (Russell 1912/1999, 40, emphasis in original). (Russell speaks here of the analysis of propositions containing descriptions, but in fact the same principle holds for all propositions.)

On this view, if we are not acquainted with entities of a given sort, then we have no direct reason to believe that such things exist. If we accept them into our ontology at all, it must be on the basis of an inference. And if we have no reason to believe that entities of a given kind exist, then it is illegitimate to use a language which contains terms which appear to refer to entities of that kind. Ontological questions are settled by seeing what

---

[7] The view I have attributed to Russell is the only account he has, during the period 1900–1918, of how we understand language. For most of that period, however, the view of language and thought of which that account is a part coexists with a different and incompatible view of language. See Levine (2009, 2016); also Hylton (forthcoming).

sorts of entities we are acquainted with, and what we can legitimately infer from knowledge based on that acquaintance.

Carnap, as I have said, seems to hold that a view of this sort is the natural alternative to the Principle of Tolerance and the rejection of ontological questions. In 'Empiricism, Semantics, and Ontology', he says that Berkeley and Hume 'denied the existence of abstract entities on the ground that immediate experience presents us only with particulars, not with universals' (219). He then goes on to say: 'Some contemporary philosophers, especially English philosophers following Bertrand Russell, think in basically similar terms. They emphasise a distinction between the data ... and the constructs based on the data. Existence or reality is ascribed only to the data; the constructs are not real entities; the corresponding linguistic expressions are merely ways of speech not actually designating anything' (219f.). (It is worth noting in this context that, in spite of this disagreement, Carnap was very much influenced by Russell's work in epistemology, perhaps especially by his *Our Knowledge of the External World*.[8] The sort of view that Carnap attributes to Russell relies upon the idea of acquaintance, or on some other conception of 'immediate data'. If views of that sort were the only alternative to the Principle of Tolerance, then Carnap's view would be very plausible. Certainly, Quine would have no sympathy at all for a view of the Russellian sort. As we shall see, however, Quine has a quite different way of arguing against the Principle of Tolerance, and has a quite different alternative.

12.7 Quine's Rejection of the Principle of Tolerance

To reject the Principle of Tolerance is to hold that the choice of language is no more a free and unconstrained choice than is the choice of theory within a language. It is to hold that the two kinds of change are made on the same general epistemological basis. The same very general terms of criticism apply both to languages and to theories: we can speak of one as better or worse than another, and of one as being the best available at a given time. Crucially, the same sorts of considerations that make one scientific theory better than another also make one scientific language better than another. That is Quine's position, insofar as he accepts the distinction between choice of language and choice of theory. But his grounds for rejecting the Principle of Tolerance are very different from the grounds on which Russell would have done so.

The mature Quine has no more sympathy for the idea of immediate or pre-theoretical access to reality than the mature Carnap has. He does not

---

[8] See Carnap (1963a), especially 13f. and 16.

hold that we first grasp some aspects of the world (as Russell thought we did by acquaintance) and then evaluate languages by seeing how accurately each reflects the world as thus revealed to us. To the contrary: he evaluates languages as he evaluates theories, by seeing how well each accommodates what has been singled out as observational evidence. (This singling out is itself guided by the overall theory, not logically prior to it; developments in theory may change our view as to what counts as evidence for theory.) In some cases, changing from one language to another enables us to formulate a theory which better accommodates the evidence. In such a case, the change of language is justified, and justified in the same very general sort of way in which the adoption of a new theory within a language is justified: it gives us a better, simpler, more efficacious way of accommodating the evidence.

For Carnap, it is crucial that the adoption of a new language be different in kind from the adoption of a new theory within a given language. In particular, the two kinds of change are not justified in the same sense. The adoption of a new theory is the ordinary kind of theoretical change, susceptible to ordinary theoretical kinds of justification. The adoption of a new language, however, is a different sort of change, made on quite different sorts of grounds. Carnap sometimes marks the distinction he draws here by calling the adoption of a new language a *practical* or *pragmatic* issue, as opposed to the theoretical matter of a change of theory within a language.

Quine's challenge is whether there is a real epistemological distinction here. If we focus on low-level aspects of theory, such as the question whether there are brick houses on a given street, to use one of Quine's examples, it may seem as if theoretical questions, which are internal to a language, are indeed quite different from those that Carnap wants to count as practical questions, which call the choice of language into question. The example of brick houses may make it seem as if there is a clear contrast between theoretical questions and questions of language choice: the former are straightforward matters of observation, whereas the latter do not seem to be answerable to observation in any very clear way. As Quine emphasises, however, matters look quite different if we focus on high-level scientific theories rather than on mundane claims about easily observable matters. For scientific theories, the considerations relevant to internal questions include whether a suggested change of theory enhances the simplicity and efficacy of the whole theory: in other words, the considerations which are relevant to a change of theory will at least sometimes include those which Carnap had called pragmatic. Hence Quine's call for 'a more thorough pragmatism' (1951f, 46). I take this to be a call for the recognition that the sorts of pragmatic considerations which Carnap thinks of as playing a role in change of language also play a role in change of theory within a language. Likewise, the sorts of questions which Carnap called external may involve change of language but are in fact made with the aim of improving our overall theory of the world. There is no epistemological distinction in kind between the two sorts of changes – and hence no reason to think that the one kind of change is a matter of convention while the other is not, and no reason to think that the one kind of change is a matter for tolerance while the other is not.

The Quinean argument which I have been briefly deploying against the Principle of Tolerance is familiar as part of his argument against the analytic–synthetic distinction. (For that reason, I have merely gestured at it, rather than spelling it out at length.[9]) Quine, as we have seen, does become somewhat more sympathetic towards the idea that in some cases a distinction can be made which deserves the name the 'analytic–synthetic distinction'. He accepts that we can, in some cases, distinguish those changes of theory which do not involve change of meaning from those changes which are also changes of meaning, and thus, strictly speaking, changes of language. But on the point that we have just been discussing – his attitude towards the Principle of Tolerance – his views never change. For this reason, Quinean analyticity is analyticity without the Principle of Tolerance.

Analyticity without Tolerance is strikingly different from Carnap's analyticity, and that of other philosophers. In the 1991 paper 'Two Dogmas in Retrospect', Quine sums up his view by saying: 'I recognize the notion of analyticity in its obvious and useful but *epistemologically insignificant* applications' (1991, 397; emphasis added). Why is Quinean analyticity 'epistemologically insignificant'? His acceptance of a notion of analyticity is an acceptance that some changes in theory are also changes in meaning. On his view, however, that does not imply that those changes are made on a different kind of epistemological basis from changes within a language. The fact that a sentence is analytic may do something to explain why we accept it – very roughly, it is a sentence that we came to accept in the course of learning to use the words it contains. But the analytic status of a sentence is not a justification of it, and does not mean that we cannot change our minds about its truth. That is a point that Carnap would accept. But for Quine, unlike Carnap, if we do change our minds about the truth of such a sentence, we do so on exactly the same very general sorts of grounds that lead us to change our minds about sentences that

---

[9] For further discussion, see Section 12.6 of the work referred to in fn. 2 above.

are not analytic. So the sort of distinction that Quine accepts between the analytic and the synthetic does not mark an epistemological cleavage: it is for this reason that Quinean analyticity, analyticity without the Tolerance, is indeed 'epistemologically insignificant'.

12.8 Ontology in Quine

To return to the issue of the status of ontology: analyticity without the Principle of Tolerance does nothing to support the idea that ontological claims are nonsensical, based on confusion, or lacking in cognitive content. So it is not surprising that Quine's work continues to contain ontological claims, such as the claim that sets exist but properties do not, even after he comes to accept a form of analyticity. What are such claims based on? Quine is famous for holding that that to be is to be the value of a variable or, as he puts it a little more cautiously in 'On What There Is: 'To be assumed as an entity is, purely and simply, to be reckoned as the value of a variable' (1948b, 13). As the more cautious version makes clear, this criterion simply tells us what a given body of theory *says* there is: we are committed to holding that those entities actually exist only to the extent that we are committed to accepting the given body of theory.

For Carnap, our commitment to a given theory is never absolute. Truth, for Carnap, is itself a relative notion – it is relative to the choice of language. A theory uses a certain language, and, according to the Principle of Tolerance, that language is freely chosen. Given the choice of that language, a given theory may be true, and hence the objects it quantifies over may exist, but this is existence relative to the choice of language, not existence in an absolute sense.

Quine, by contrast, does not accept the Principle of Tolerance, and hence does not accept that truth and ontology are language-relative. We adopt the best language-cum-theory that we can; what exists are the entities that that theory quantifies over. Whether, or to what extent, we can separate the adoption of a language from the adoption of a theory within the language may be crucial for the question of analyticity but for our present purposes, for the question of the status of ontology, it does not matter. Even if the separation of the two kinds of steps were always clear-cut, which Quine does not accept, it would make no difference to this point. The reason is that the two kinds of steps are taken on the same very general sort of basis, and they both aim at the same goal: our having the best language-cum-theory that we can come up with. Our view as to which is the best may well change, and the ontology that we accept may change along with it, but while we hold a certain language-cum-theory we count it as true and count the objects that it quantifies over as existing – really existing, not merely existing relative to some choice of language. This is what gives sense to a question such as whether there are properties: we evaluate whether the best available language-cum-theory would quantify over properties.

Carnap's view that ontological questions are pseudo-questions depends on the analytic–synthetic distinction, as we saw in Section 12.4. But our recent reflections show that it also depends on the Principle of Tolerance. For that reason, Quine never accepts Carnap's view of the status of ontology, even when he comes to be somewhat sympathetic to the view that there is a distinction to be made between the analytic and the synthetic.

CHAPTER 13

Carnap, Quine, and Williamson
Metaphysics, Semantics, and Science

Gary Kemp

For all their much-discussed disagreements over analyticity and ontology, Quine shared Carnap's more fundamental commitment to 'scientific philosophy'. The primary service of legitimate philosophy is to clarify, to make more precise, and to make more explicit the methods and deliverances of science. Sometimes the philosopher may extend those methods to novel areas – as in Quine's description of radical translation or Carnap's of semantics and modality – but decidedly there remains no room for metaphysics in some extra-scientific sense.

The mature development in Quine was what he called his 'naturalism', the view that 'it is only within science, and not in some prior philosophy, that reality is to be identified and described' (1981d, 21). For his part, Carnap wrote that the 'logic of science ... is in the process of cutting itself loose from philosophy and becoming a properly scientific field, where all work is done according to strict scientific methods and not by means of "higher" or "deeper" insights' (1934/1937, 46; evidently, in a sign of the times, Carnap intends 'philosophy' pejoratively, and that his work is properly scientific). Of course, there are well-known difficulties in calling Carnap a naturalist in Quine's sense. For my purposes this issue can be avoided, but one difference that will figure later is that unlike Quine, Carnap thinks of the proper task of the philosopher as that of designing entire symbolic languages for science, and of philosophical communication in the vernacular as but a practically necessary propaedeutic to the expression of maximally objective, interest-free theoretical truths in a symbolic language. Roughly speaking, whereas for Carnap pragmatic matters and theoretical matters are mutually exclusive, for Quine *all* matters are both pragmatic and theoretical. Nevertheless, Carnap and Quine are agreed that many notions of common sense and many questions couched in ordinary language, as well as the many untethered doctrines of the speculative thinker, are simply not suitable as they stand for serious philosophical theories or serious philosophical examination. Philosophical questions themselves must be shown to arise within science if they are genuinely to be factual ones.

In this essay I compare this shared ground between Quine's view and Carnap's with a recent estimate of the role of philosophy vis-à-vis science, which might be summed up as the re-enshrinement of metaphysics as sharing the stage with science: that advanced by Timothy Williamson (2007, in his book *The Philosophy of Philosophy*).[1] Now Williamson rejects both Quinean naturalism (1–2) and Carnapian positivism (51), but he does not think of his view as an explicit response to Quine or Carnap, and more generally he does not think of their views as paradigmatic of the 'linguistic philosophy' he explicitly rejects – he thinks rather of such figures as A. J. Ayer, Peter Strawson, and Michael Dummett. There is nevertheless a basic disagreement between Williamson, on the one hand, and Carnap and Quine, on the other, one that will require further understanding of the point gestured at in the last paragraph to appreciate. The disagreement is partly about science generally and partly about semantics in particular. The upshot is that from either Carnap's point of view or Quine's, this recent ascendance of metaphysics – partly abetted by Williamson – will seem an unjustified departure from science.

13.1 Carnap and Quine: Some Similarities and a Difference

All scientists properly so-called strive among other things for maximum clarity – ideally, for their products to be capable of formalisation, even if in most cases it would be pointless actually to carry it out. But neither Quine nor Carnap assume a narrow view of science: they intend 'science' in the broader sense of *Wissenschaft* (The term is partly an honorific). We have a cluster of sciences, the centre being mathematics, then physics and chemistry, then geology, biology, psychology, cognitive science, social science, linguistics, economics, history, anthropology, politics, religious studies, and the like, as we move towards the outskirts. We can form several rough gradations – which sometimes coincide and sometimes do not – of precision, objectivity, generality, and the acceptance of general laws; of freedom from being infected with practical interests, story-telling, metaphor, and rhetoric; and of methodological disputes – relatively absent at the one end and ubiquitous at the other. Nevertheless, Carnap and Quine both value

---

[1] There is also Williamson's less substantive 'How Did We Get Here from There? The Transformation of Analytic Philosophy' (2014). *Modal Logic as Metaphysics* (2015) contains more dimensions of his view, but space here is limited.

the formal expression of science, at least for those with philosophical interests. Carnap famously devises new artificial languages for the purpose, as in *The Logical Syntax of Language* (1934), whereas Quine speaks of 'regimentation', as in *Word and Object* (1960), and partly practises it.

For both Carnap and Quine, ordinary language – the kind used to express common-sense ideas, or 'folk theories' – is often decidedly unsuitable for the expression of science proper. Consider Zeno's paradoxes. It took a colossal amount of time and effort until, in the nineteenth century, mathematics and physics could give satisfying answers to these ancient riddles; this required much painstaking clarification and rigorisation of certain conceptual implements – differentiation, integration, the limit, and so on – on the part of Newton, Cauchy, Weierstrass, and many others. 'Ordinary language is only loosely factual' (2008a, 285), said Quine, and no wonder; it arose amidst the traffic of ordinary people with ordinary needs, and it would be strange if it had more determinacy than such needs called for. Commenting on the conundrums of personal identity, Quine writes: 'They are questions about the concept of person, or the word "person," which, like most words, goes vague in contexts where it has not been needed. When need does arise in hitherto unneeded contexts, we adopt a convention, or receive a disguised one from the Supreme Court'. (Quine 1995a, 39) Quine and Carnap agree that such conceptual engineering is not happily characterised as 'analysis' – with the implication of finding out what was there in language all along, as suggested by the analogy with chemical analysis – but instead should be characterised as 'explication' in their special sense: the replacement of existing words or phrases with more precise ones, normally homophonic with the old ones but not necessarily. Where what is in question is a not-yet-explicated φ, it is often pointless to insist that there is an essence of φ to be drawn out, either by 'conceptual analysis' of the term for φ or inquiry into the 'nature' of φ.[2] Still, the notion may be judged to be scientifically important, governed by some principle or conjunction of principles, even if the notion is crucially vague, and typically will seem to invoke unwanted entities or entities one considers superfluous. The task then is to replace the notion, validating the principles in terms of entities which already have the seal of scientific approval.

Quine is, however, much more inclined than Carnap to stress the commonality between ordinary language and formal language. Quine avows 'my attitude toward "formal" languages is very different from Carnap's.

---

[2] In *Logical Foundations of Probability*, Carnap gives a good example: is the workaday concept of probability 'degree of confirmation' or 'frequency'? There is no real issue here, in Carnap's view (23–9).

Serious artificial notations, e.g. in mathematics or in your logic or mine, I consider supplementary but integral parts of natural language' (Quine's letter of 1943 to Church; Quine and Carnap 1991). He conceives the latter as an 'outgrowth' of the former, achieved by paraphrase of troublesome parts of the latter, not generally by outright replacement of whole languages, as Carnap favours. Carnap's task rather was to construct formal languages from the ground up, as in Languages I and II of *The Logical Syntax of Language*. This is indeed a symptom of a deep disagreement between the two, yet for these purposes we shall soon find their two paths reuniting at a still deeper level.[3]

Although he formulated it comparatively late (1950a), Carnap's distinction between 'external questions' and 'internal questions' can retrospectively be seen as central to his view from around 1931 onwards. An internal question is one posed within a 'linguistic framework' – an artificial language – and admits of an answer according to two types of general principles operating within that language: (1) 'transformation rules', including primitive sentences (axioms, L-rules of 1934/1937, §§10–14, 30–3) and rules specifying a logical consequence relation; and (2) – assuming it is a language set up for empirical enquiry – 'concrete sentences' (or 'protocol sentences') registering empirical data, plus 'theoretical sentences' (P-rules of 1934/1937, §51). The result is that the truth-value of an internal sentence depends on certain observations irrespective of whether it is concrete or theoretical, unless the sentence is analytic (L-true) or contradictory (L-false). By contrast, external questions are posed from outside the linguistic framework or artificial language. Such a question does not have a fully objective and precise answer, does not have a strictly rational answer, and indeed cannot genuinely have a true or correct answer. What it does have is a practical answer, an answer which involves the aims and desires of the investigators. For example, the ontological question 'Do numbers really exist?' cannot have the status that the metaphysician envisages. As an internal question, it admits of a trivial, deflationary answer, for either an artificial language will or will not have number terms and the answer will be analytic. As an external question, the question is misleading; it is more perspicuously posed as something like 'Is this language suitable for conducting mathematics?' Not only will the answer come in terms of less and more, it cannot strictly be assigned an objective status.

---

[3] Quine: 'In later years [Carnap's] views went on evolving and so did mine, in divergent ways. But *even where we disagreed he was still setting the theme*; the line of my thought was largely determined by problems that I felt his position presented' (Quine, 1970a, 41; emphasis added).

The cash value of the internal–external distinction of 1950 was prefigured in the so-called Principle of Tolerance, which appeared in *The Logical Syntax of Language*:

> It is not our business to set up prohibitions, but to arrive at conventions ... In logic there are no morals. Everyone is at liberty to build up his own logic, i.e. his own language, as he wishes. All that is required of him is that, if he wishes to discuss it, he must state his methods clearly, and give syntactical rules instead of philosophical arguments. (Carnap 1934/1937, 51–2).

The basis for the Principle is, however, made more precise and explicit by the internal–external distinction. According to that distinction, because logical relations do not determinately bite except within an artificial language, one cannot subject a person to purely rational criticism for setting up whatever artificial language the person likes; one should be tolerant of such people, however little one may share their interests.

Carnap invokes the internal–external distinction in connection with ontology, but the idea can be seen to apply more generally to logic, and in particular to analyticity. For the distinction has a corresponding deflationary effect on questions of analyticity: just as answers to questions of ontology are settled by the choice of language, so it is with questions of analyticity. Carnap takes analytic truths primarily as being not discovered but stipulated, as being made true by their being decreed as such in the very setting up of the language (see Quine and Carnap 1991, 427f). Although he does maintain that the idea makes sense with respect to sentences of ordinary language, there is no precise fact about which sentences of ordinary language have this status.[4] '[T]he concept of analyticity has an exact definition only in the case of a language system, namely a system of semantical rules, not in the case of an ordinary language, because in the latter the words have no clearly defined meaning', he writes (1952/1990, 427). The ordinary concept is 'vague and ambiguous, and basically incomprehensible' (1963a, 919; see also 1956 234, 240, 241; and Quine and Carnap 1991, 430). Thus, it is ultimately futile to debate the question of the analyticity of a sentence of ordinary language, even if there are manifestly central examples.[5]

---

[4] For Carnap the discipline called 'descriptive' or 'empirical' semantics is well grounded. But Carnap holds that the concepts of that discipline will often have a region between 'applies' and 'the negation applies'; hence the need for artificial languages.

[5] Sam Hillier writes: 'In "Testability and Meaning", Carnap also makes it clear that his syntactical stipulations are intended to function as models of the language of science ... he remarks that "there is no sharp line between observable and non-observable predicates ... for the sake of simplicity we will here draw a sharp distinction [...]" (1936/7,455) ... The same holds true for his discussion of the analytic/synthetic distinction' (2010).

Carnap tirelessly maintained and defended the analytic–synthetic distinction despite Quine's qualms over it, which culminated in Quine's having explicitly rejected it in 'Two Dogmas of Empiricism'. Quine in effect asks: when you, Carnap, are stipulating a certain class of sentences as 'analytic', what exactly is it that you are stipulating? Quine thought of Carnap's seeming failure to give a satisfactory answer to his question as showing a fundamental weakness in his view concerning the precise role of artificial languages. It is not satisfactory to explicate the notion of analyticity simply by stipulating an extension as Carnap suggested; that doesn't answer the question of what it is that is being stipulated.

Yet in later work – in *Roots of Reference* (1974a) and 'Two Dogmas in Retrospect' (1991) – Quine accepted a version of the analytic–synthetic distinction. This later definition locates it squarely in the realm of the factual, indeed of the behavioural. It comes in two stages: (i) a sentence is analytic for a person if and only if the person 'learned the truth of the sentence by learning to use one or more of its words' (1991, 395–6); and (ii) a sentence is simply analytic if and only if it is analytic for those who are competent with the relevant terms.[6] This is close to what is nowadays called 'Epistemic Analyticity' – that a sentence is analytic if a linguistically competent person cannot understand the sentence without accepting it – the differences being that this notion of Epistemic Analyticity lacks the reference to 'learning' in Quine's definition, using instead the term 'understanding' roughly in place of 'learning', and that it lacks references to a particular speaker, as in Quine's definition (I set aside the thought that Epistemic Analyticity is more explicitly normative than Quine's idea).

Quine supposes further that analyticity is preserved by 'obvious' logical inference: a sentence that is reached in logically obvious steps from an analytic sentence is itself analytic (1974a, 78–80; 1991, 396). Granted that a complete set of axioms may be chosen which are analytic in his sense, he thus allows that logic is analytic. Peter Hylton puts this together with another point from 'Two Dogmas', of the falsity from Quine's point of view of the Principle of Tolerance, of the inextricability of the pragmatic from the logical: the rational justification for *any* sentence has both theoretic factors and pragmatic factors (2019, 14–16). The result is that decisions about whether to accept a sentence are not rendered moot by the analyticity of the sentence. It might rationally be discarded despite being

---

[6] The Quine 1974a treatment is not quite as forthright as this, and includes the word 'understanding' in the discussion; see 'Cognitive Meaning' (1979) for Quine's attempt to find a behavioural definition of the word.

analytic (as some think the principle of bivalence should be, in light of quantum mechanics). Thus, the status of a sentence as analytic is not 'epistemically significant'; it is only of academic interest that a given sentence is analytic. Quine was surely motivated, at least in part, by the desire to find a sense of 'analyticity' that would make something like the customary verbal distinctions, yet without disturbing his epistemological picture in any way.

Hylton's point also shows that despite this disagreement between Carnap and Quine, they remain largely in agreement on a key point about ordinary knowledge, concerning assertions of natural language. For both Carnap and Quine take a critical view of ordinary language; neither thinks that the analyticity of a sentence of ordinary language is of great pith and moment. Quine writes the following regarding ontology, but the point goes over easily into the status of unregimented ordinary language in its logical or 'ideological' aspect:

> The common man's ontology is vague and untidy in two ways. It takes in many purported objects that are vaguely or inadequately defined ... [And] we cannot even tell in general which of these vague things to ascribe to a man's ontology ... where can we draw the line?
>
> It is a wrong question; there is no line to draw. Bodies are assumed, yes; they are the things, first and foremost. Beyond them there is a succession of dwindling analogies ...
>
> My point is not that ordinary language is slipshod, slipshod though it be. We must recognize this grading off for what it is, and recognize that a fenced ontology is just not implicit in ordinary language. The idea of a boundary between being and nonbeing is a philosophical idea, an idea of technical science in a broad sense. Scientists and philosophers seek a comprehensive system of the world, and one that is oriented to reference even more squarely and utterly than ordinary language. Ontological concern is not a correction of a lay thought and practice; it is foreign to the lay culture, though an outgrowth of it. (1981d, 9)

Many ordinary expressions are simply not suitable for the expression of 'technical science in a broad sense'. This implies than many questions posed in ordinary language, many curiosities, should be rejected as naïve or confused by the light of science.

Carnap, for his part, took Heidegger to task for being unaware of the advances in quantification theory made by Frege and others (1932/1959, 69). Heidegger uses, for example, the term 'nothing' as if it were name of an entity. But according to modern logicians, 'nothing' is best construed as a quantifier, not as a referring term; there isn't any object into whose nature one might inquire (on the standard 'syncategorematic' explanation of quantifiers). Sentences such as Heidegger's 'What about this Nothing?' are such that in a logically well-behaved language, no such sentences 'can even be constructed' (70). Such are the traps of using the ancestral lore. The same goes for what he terms 'metaphysical' questions in general; they are, in general, not strictly clear if posed in terms of the unreconstructed vernacular.

Something similar goes for Sorites-inducing expressions such as 'heap', 'big', or 'dry': though such terms are fine for everyday use, Sorites-type puzzles arise for them (as well as Ship-of-Theseus puzzles for artefactual terms – not to mention the complexities of 'person', where many urgent and often competing interests converge). It is not by chance that these kinds of paradox are seldom felt in real life; for where they do arise, folk may adjust their usage of the relevant term to avoid the paradox or uncertainty. For purposes of science (but also sometimes for other purposes, such as legal purposes), the way to escape the miasma is to replace such terms, to explicate them, or simply not to use them – for example, to dispense with the monadic term 'big' and make do with the relational term 'bigger than'. Thus, Carnap thinks it's often useless work to *analyse* existing concepts or things, and instead recommends 'rational reconstruction' (Carnap 1928/1967).

13.2 Williamson

Since I am concerned only with points of contrast between his view and the shared regions I have outlined between Carnap's and Quine's, there is much in Williamson's view that I will not discuss. For example, I am not going to discuss at any length the role (or the nature) of modal concepts – so central to Williamson's philosophy and seemingly discounted by Quine's philosophy if not by Carnap's. I shall simply pick out certain interesting parts of *The Philosophy of Philosophy* which express his view in a revealing way when compared with aspects of the Carnap-Quine view that I have outlined. It may indeed be a case of quoting him 'out of context', but I hope not unfairly. I'll concentrate first on Williamson's denial that philosophy is primarily a linguistic activity, and second on his related estimate of philosophical thought experiments.

13.2.1 *Science, Semantics, and the Linguistic Turn*

Williamson inveighs against 'philosophical exceptionalism', apparently approving of Quine's disavowal of 'first philosophy' in his opposition to the idea of 'metaphilosophy' (2007, Preface). In his opinion, the methods of philosophy are not categorically different in kind from those of science, even if it goes in much more for the armchair and much less for the laboratory. And like Carnap and Quine, Williamson discounts the purported special relevance of 'intuitions' to philosophy. They are just judgements, which can be repudiated like any other judgement by evidence or argument.

For Williamson, however, there are distinctively philosophical questions which are largely left unaddressed by scientists of the lab-coat variety. Partly these are metaphysical questions – questions of essence, of necessity, of the nature of objects and properties, of ontology and existence, and so on. And partly they are epistemological questions – the nature and the extent of knowledge, belief, and justification, the role of meaning and concepts in knowledge, and so on. For Williamson, it is an extraordinary rich field of problems calling for the sharpest logical implements and minds to go with them to make genuine progress.

Carnap and Quine would squirm at this. In their estimation, these words – 'essence', 'property', 'meaning', 'belief', 'knowledge', 'concept', and so on – are just words of ordinary usage.[7] Asking point-blank questions of the form 'What is the nature of …?' may mislead one into the mere assumption that science should retain the given category, or that there must be interesting *de re* modal truths about the category that require special philosophical techniques to bring out. On the contrary, descriptions of things in these terms are often imprecise or vague, relative to our shifting and unpredictable interests, and therefore not suited for science (the recent run of results investigating the 'method of cases' in 'experimental philosophy' comport with this; Stich and Tobia 2018). Of the term 'knowledge', Quine avers that what the long-running industry begun by Gettier has shown is that 'knowledge is a bad job'; that we should simply 'make do with its separate ingredients' (1989, 109; see also 2008a, 176, 249, 322[8]). In normal circumstances, the inchoate theory surrounding the word makes for a well-established and profitable use, but in abnormal circumstances it

---

[7] In his response to Schuldenfrei, Quine writes: 'My position is that the notions of thought and belief are very worthy objects of philosophical and scientific clarification and analysis, and that they are in equal measure very ill suited for use as instruments of philosophical and scientific clarification and analysis' (Quine 1981b, 185).

[8] Quine: The '[d]efinition … of "knowledge" is in trouble since Gettier's challenge of the definition of knowledge as true and warranted belief' ('The Innate Foundational Endowments', 2008a, 176); 'Knowledge, nearly enough, is true belief on strong evidence. How strong? There is no significant cut off point. "Know" is like "big": useful and unobjectionable in the vernacular where we acquiesce in vagueness, but unsuited to technical use because of lacking a precise boundary. Epistemology, or the theory of knowledge, blushes for its name' (2008a, 322).

may fail, like a Volkswagen Beetle being asked to climb a steep and muddy slope off road. Such terms arose in the context of human beings with specific aims, such as cooperation and solving workaday practical problems, ones which seldom call for scientific precision.

Let us now compare Williamson's attitude towards analyticity with Quine's and then with Carnap's. Williamson finds that neither analyticity in the 'epistemological' sense nor analyticity in the 'metaphysical' sense can make good on the traditional idea that analytic truths are categorically unlike other truths, insubstantial or 'trifling' in Locke's sense. '[P]hilosophical truths are analytic at most in senses too weak [not sufficiently distinctive] to be of much explanatory value or to justify conceiving contemporary philosophy in terms of a linguistic or conceptual turn' (53), he concludes. But this also is true of Quinean analyticity, which as we have seen includes logic and perhaps a sizeable proportion of mathematics. As recently noted, Quine recognises a sense of 'analyticity' which is very close to the epistemic sense, but he does not claim that that sense of analyticity is therefore epistemologically significant. Quine therefore more or less agrees with Williamson on this point.

The situation is more delicate with respect to Carnapian analyticity. Carnap, as explained above, does not think that questions of whether or not a sentence of ordinary language is analytic are worth agonising about. Not only can one quibble until the cows come home about 'whatever has shape has size', 'force = mass times acceleration', or even 'bachelors are unmarried', philosophical semantics since the 1970s have made it seem that 'dogs are animals' and the like – which many people had assumed to be analytic – are *not* true-by-meaning, so not analytic, for a term such as 'dog' is held to be a natural kind term, and it is conceivable, though unlikely, that we are mistaken about the nature of their referents. For Carnap these cases cry out for explication. In setting up an artificial language, these are all questions that one is free to make decisions about. Thus, if what is in question is Carnap's attitude towards the concept of analyticity with respect to natural language, then he assuredly does not propose to conduct epistemology armed with that concept. If on the contrary what is in question is Carnap's explicated, artificial-language version of analyticity, then he certainly *does* think that the concept is epistemologically significant, but it is unclear that Williamson has a criticism of Carnap's plan for artificial languages, even if the plan were judged not to be, as Carnap believed, philosophically central.

So analyticity is not the place to locate a substantial difference between Quine and Carnap, on the one hand, and Williamson, on the other.

Nevertheless, they do substantially differ from Williamson over empirical semantics, with extensive repercussions for other topics. For Williamson's attitude towards certain other key words of semantics itself is apt to be puzzling from the Quine–Carnap explicative point of view. Of the term 'synonymous', he writes:

> Although [Quine] may succeed in showing that "analytic" is caught in a circle with other semantic terms, such as "synonymous," he does not adequately motivate his jump from that point to the conclusion that the terms in the circle all lack scientific respectability, as opposed to the contrary conclusion that they all have it. Given any science, someone may insist that it define its terms, and the terms used to define them, and so on until it is driven round in a circle. By itself, that hardly demonstrates the illegitimacy of the science. Every discipline must use undefined terms somewhere or other ... After all, semantics is now a thriving branch of empirical linguistics. (50)

And: 'Pairs such as "furze" and "gorse" are pre-theoretically plausible cases of synonymous expressions that speakers can understand in the ordinary way without being in a position to know them to be synonymous' (67).

We can grant that these are 'pre-theoretically plausible cases of synonymous expressions', but are they *really* synonymous? The uncertainty may be with the term 'synonymous'. The common-sense use of the term is interesting, but a well-grounded verdict depends on how the word comes to be satisfactorily defined or whether indeed such a word is retained in a mature theory. Perhaps 'semantics is now a thriving branch of empirical linguistics', but the standing of the field does not depend on whether there is a well-grounded and fruitful definition of 'synonymy' to be had, as Quine repeatedly emphasised.

The circle of semantical terms of which Williamson speaks is by no means sufficiently stable to be generally accepted as is by working scientists. One can imagine various senses of what it means to be 'thriving', but at any rate the sub-region of linguistics called 'semantics' is more towards the ubiquitous end of the rough scale of sciences described at the outset of Section 13.1. There are just too many views about what are its proper methods, aims, and concepts to declare the existence of anything like a dominant paradigm – there is teleosemantics, truth-maker semantics, inferentialism, conceptual role semantics, situation semantics, game-theoretic semantics, dynamic semantics, intention-based semantics, semantical minimalism, expressivist semantics, radical contextualism, and others. And as Chomsky himself warns, many of these might well incorporate philosophical speculation masquerading as science (2000, 40–5).

Williamson says, 'When clarification is needed in some specific respect, it can be achieved by stipulation or otherwise, as elsewhere in science' (51), but he does not actually propose what for Carnap and Quine would be a viable explication of the term 'synonymy' (as applied to natural language). This verges on the view that the ordinary concept of synonymy is sufficiently shipshape just as it stands. But from Carnap's or Quine's point of view, it is unscientific to presuppose a pre-existing distinction between problems of meaning and problems of fact, implying the possibility of a sharp and well-grounded notion of synonymy. It is not mere scepticism to suspect that the term is no more than a practical device of ordinary language, not a term to rest with in science itself.

Another telling aspect of Williamson's view emerges in connection with the problem of vagueness:

> Some philosophers ... regard the attempt to give a systematic statement of the truth conditions of English sentences in terms of the meanings of their constituents as vain. For them, the formalization of "Mars was always either dry or not dry" as $(\forall t) (Dry(m, t) \lor \neg Dry(m, t))$ is already a mistake. This attitude suggests a premature and slightly facile pessimism. No doubt formal semantics has not described any natural language with perfect accuracy; what has not been made plausible is that it provides no deep insights into natural languages. (37)

A reasonable move from the Carnap–Quine point of view would indeed be to 'deny the relevance of formal semantic theories to vague natural languages' (37; not to dispute Williamson's criticisms of the other formal approaches to vagueness, such as supervaluation or degrees of truth). Lay people may well regard the setting up of premises for a Sorites-style reduction to absurdity as setting them up for a trick, a trick moreover that does not strike them as profound, the reaction being a shrug of the shoulders or an eye roll. When confronted explicitly with the induction premise – for example, that if x is not dry then x with a single molecule of liquid removed is not dry – they will perhaps agree with Peter Strawson that 'ordinary language has no exact logic' (Strawson 1950, 344), even though it may be unclear to them exactly what they are denying. Words such as 'dry' are only as precise as ordinary purposes demand. For extraordinary purposes, for scientific purposes, we can replace it with 'dryer than' (as Carnap replaces 'warmth' by 'temperature'; Carnap 1950/1956, 29). One simply refuses to play the trickster's game.

Neither Carnap nor Quine took the linguistic turn in the sense Williamson intends, yet their views about semantics and science render their general philosophical views manifestly antithetical to Williamson's.

Carnap's interest in rational reconstruction and Quine's in regimentation place the language of science at the centre of philosophical concern, but for this purpose they avoid using the intuitive semantical notions in effect favoured by Williamson and others to describe natural or ordinary language. They use only at most their corresponding explicated versions, and strictly only with respect to artificial or regimented languages.[9] Neither Carnap nor Quine thinks that there is a special realm of philosophical truth to be divined by reflecting on our ongoing and sometimes unrestrained thought and talk about the world, as opposed to clarifying, streamlining, and marginally adding to our best scientific theories of the world.

13.2.2 Thought Experiments and Science

I now shift to my other point of comparison, that of their respective attitudes towards 'thought experiments'. I shall argue that – from the Carnap–Quine point of view – Williamson is wrong to deny that there is a fundamental difference between thought experiments in philosophy and those in empirical sciences. I shall begin with some points about the subject in general.[10]

There is no doubt that the rhetorical, heuristic, and pedagogical value of thought experiments is often striking – no less than Galileo, Newton, and Einstein provide examples. They can serve to dramatise aspects of the world, to isolate and highlight them by descending from the general to the particular level, even if the particulars are hypothetical and often ideal. It seems natural to divide thought experiments into two kinds, which naïvely we might call 'empirical' and 'philosophical' (or 'conceptual'); Williamson questions that division, at least that there is an epistemologically relevant difference.

A typical laboratory experiment involves a theory that implies that certain things will happen under certain circumstances. One brings about those circumstances and checks for those things. The motive is that one is not sufficiently certain what result will be obtained, or that one needs the result to persuade others, or both; for these reasons one must get out of one's armchair – a mere thought experiment is unavailing.

Where one does, in a certain sense – perhaps the sense in which the slave in Plato's Meno 'knew' geometry – already 'possess' the result a thought

---

[9] Tarski is the source of both the later Carnap's use of semantical concepts and of the post-youth Quine's. But Tarski is clear that the notions he employs are explicated versions of the (inconsistent) ordinary notions.

[10] Carnap doesn't explicitly address thought experiments I far as I'm aware; for Quine, see (1981b, 185; 2008a, 284, 341, 345, 437, 448).

experiment seems to be both appropriate and valuable. Perhaps the result is merely implicit in principles which one accepts, or one simply has not thought through the relevant application to reality of the principles. For example, Galileo considered Aristotle's claim that the speed at which a body falls is proportional to its weight, and contested it by reasoning along the following lines: if heavy things fall faster than light things, then if a small lead shot were tied to a cannonball with a length of twine and they were dropped, the cannonball's flight must be slowed as it pulls the small lead shot downwards by the twine; but on the other hand since the cannonball-and-lead-shot-assembly is heavier than the cannonball itself, it must fall faster than the cannonball falls alone; the contradiction shows the hypothesis to be incorrect, so it is not the case that heavy things fall faster than light things. We want to say that this sort of thought experiment is one of physics – is 'empirical' – despite its having been conducted from the armchair.

A well-known thought experiment which we are naïvely inclined to say is of another type, the philosophical type, was described by Gettier and is stressed by Williamson, a variant of which might run as follows: if knowledge is justified true belief, then someone who justifiably believes that there is a sheep in the field – but only because they see in the field a dog disguised as a sheep – when in fact there is sheep in the field which, however, the person does not see, would have a justified true belief that there is a sheep in the field, and therefore would know that there a sheep in the field; but such a person would not know that there is sheep in the field, so it is not the case that knowledge is justified true belief.

What is the distinction between these kinds of thought experiment? One might think it crucial that the first experiment presumably involves the physical-spatial imagination, whereas the second one does not. Williamson, however, is sceptical that the second one really lacks this feature; discussing knowledge-involving scenarios like the second thought experiment above, he writes:

> individual differences in the skill with which concepts are applied depend constitutively, not just causally, on past experience …
>
> In a similar way, past experiences of spatial and temporal properties may play a role in skilful mathematical "intuition" that is not directly evidential but far exceeds what is needed to acquire the relevant mathematical concepts. (168–9)

Thus, Williamson denies that the distinction between a priori and a posteriori knowledge captures a distinction between our two examples in 'a deeper theoretical analysis' (169).

All the same, even if it is not precisely an exercise of the a priori faculties, one might think of Gettier as making a point about the *concept* of knowledge, namely that the concept is not that of justified true belief, whereas Galileo's lesson affects the *theory* of gravity, without disturbing the concept of gravity itself. Call this reaction (a). One can also feel a tug in the opposite direction, either supposing (b) that despite appearances, and as in the case of Galileo and the concept of gravity, Gettier's conclusion does not teach us anything about the concept of knowledge so much as having implications for theories in which the concept figures; or (c) that in both cases the lesson affects the concepts themselves, for the concepts depend on the theories in which they figure.

Williamson chooses (b). Together with the theory and conditions, in both cases the result is arrived at via some subtle use of the imagination and perhaps knowledge of other matters. He argues impressively against (a) – in sum, the attempt to make out exactly why one experiment is conceptual and the other is not simply runs into the weeds (I will not go into the detail of the discussion); (c) is ruled out for similar reasons, and a further reason against it comes by considering communication (in their 1992 Fodor and LePore make a similar point):

> Such variation ... [differing results from thought experiments] may result from cross-cultural variation in the meaning of "know" or other epistemological terms, but it need not. It may occur between sub-communities of English speakers who all use the words as part of a single common vocabulary, but disagree in their applications of them, just as different communities may disagree in their applications of the word "justice" while still using it with a single shared meaning. (190)

The conclusion is that the two thought experiments are not fundamentally different.

I'll restrict myself to Carnap in this paragraph. It is doubtful that Carnap would have set much store by the idea that despite disagreement over cases, we communicate with 'the word "justice" while still using it with a single shared meaning'. But more generally I think that Carnap's view would be that the distinctions on which Williamson's discussion depends – especially that between the conceptual and the factual as applied to ordinary if learned speech but in other applications as well – are not sufficiently precise to ground any such sweeping conclusions. If the distinction coincides with the a priori–a posteriori distinction, then the a priori does not have anything like a determinate extension if the field is ordinary assertions, sentences, or beliefs. One has to explicate it in terms of analyticity, and as we've seen analyticity is only genuinely fruitful with respect to artificial languages, in Carnap's view. A question such as 'Is the reasoning [in a given thought experiment] a priori?' is often a 'pseudo-question' without cognitive content or at any rate with insufficient cognitive content, leading only to 'sterile debates'.

In Gettier cases, Putnam's twin-earth cases, and so on, there is scant possibility of the thought experiment conflicting with an actual experiment. Perhaps for that very reason, one supposes that these thought experiments have a type of authoritative standing that the physical ones lack. Yet the results are disputable. Williamson writes:

> [N]ative English speakers sometimes dispute the Gettier verdict ... In doing so, they show poor epistemological judgment but not linguistic incompetence ... We assent to (3*) ['If there were an instance of the Gettier case, it would be an instance of justified true belief without knowledge'] on the basis of an offline application of our ability to classify people around us as knowing various truths or as ignorant of them ... That classificatory ability goes far beyond mere linguistic understanding of "know." (188)

This reference to a fine-honed skill, this 'classificatory ability', is important to Williamson's case. He goes on: '[S]omeone with a distorted epistemological outlook may reject (3*) yet still possess the relevant concepts: they genuinely believe that the subject of the Gettier case would not have justified true belief without knowledge' (189).

And further:

> Much of the evidence for cross-cultural variation in judgments on thought experiments concerns verdicts by people without philosophical training. Yet philosophy students have to learn how to apply general concepts to specific examples with careful attention to the relevant subtleties, just as law students have to learn how to analyse hypothetical cases. Levels of disagreement over thought experiments seem to be significantly lower among fully trained philosophers than among novices. (191)

The reliance on this 'classificatory ability', this skill for 'applying epistemological concepts', indeed likens the epistemologist to a legal judge if not a seer, called upon to judge even in the absence of intersubjectively decisive reasons. Yet a principal reason that judges are essential to the legal system, and cannot typically be replaced by computers, is that the fact that *they judged* as they did is *itself* a central fact of interest. A proper scientist, on the contrary, tries in principle to eliminate any reliance on the judgements and skills of human beings, and in principle reckons a failure to do this a failure of the experimental design. To claim that sort of status for 'applying epistemological concepts' would require the formulation of criteria for their application – rules or methods of measurement that a computer could follow – which is exactly what Williamson's reliance on this 'classificatory ability', the idea of knowledge as primitive, rules out.

Galileo could rely on his thought experiment because he was confident of what the relevant observations of an actual experiment would be: the objects, aside from considerations of air friction, would fall at the same rate. It is not a logical contradiction that drives the thought experiment; perhaps the cannonball-and-lead-shot system behaves as a single object the more rigid the connection between them, and only minimally when held together with a piece of twine. If there were any iota of doubt, then in principle he could have rigged up an actual experiment and measured. There is no such contrast in the case of the Gettier thought experiment – there is nothing to measure. The same point can be made by observing that widespread and well-informed disagreement over the result of a Gettier thought experiment needn't be rationally resolvable, whereas widespread disagreement over a Galileo-style thought experiment would be rationally resolvable by conducting an actual experiment. True, it is sometimes practically impossible to perform an actual experiment of the requisite sort – Einstein's scenario of the observations one would make travelling at the speed of light might be cited – but still the facts being pointed out have nothing to do with the observer.

Carnap would not want to join Williamson in his evident confidence in making the distinction between cases of factual disagreement and cases of linguistic or conceptual disagreement. He would take note of the lack of unanimity of response, the sheer range and diversity of theories of knowledge, the perplexing chaos of competing accounts of what it is to know, and take this as a sign of a lacuna in our theoretical landscape. *If* there is some theoretical need to do so, then one can propose explications incorporating, for example, the causal theory of knowledge, or reliablism, or something else. From Carnap's point of view, these are simply alternative explications of the familiar word 'knowledge'. Perhaps they can *all* be accepted, with differing subscripts on the term 'knowledge'. But there needn't be a fact which can be discovered that such a response brings to light, one which was already there, hidden somehow from sight.

The situation is similar for such questions as 'Do cats understand?', 'Do chess-robots think?', or 'Why do things happen?', to borrow examples from Chomsky (2000, 19, 45, 114). The feeling that such questions are well formed and have answers just as they stand is sometimes undeniable – and is encouraged no doubt by the thought, so popular since Putnam and Kripke discussed the topic, that ordinary words stand for natural kinds – but one can join with Chomsky (and Wittgenstein) in suggesting that these, like the question of whether submarines swim or whether Pluto is a planet, are at best matters of decision, not fact (even if some decisions might fit emerging science better than others, such as the recent decision to exclude Pluto from planethood).

I have so far discussed Williamson's view of thought experiments without discussing its most significant feature, that at their centre is counterfactual reasoning (involving conditionals of the form 'If P were so, then Q would be'; 141ff). Space permits to make only a point concerning that arch anti-modalist Quine.

Quine observed that many counterfactual conditionals have a hidden egocentric element, as in the pair of examples that he cites from Nelson Goodman: 'If Caesar were in command, he would use the atom bomb; If Caesar were in command, he would use catapults' (1960, 222). It is only a specific context, in particular a speaker intention, that determines which of these sentences is appropriate (a supposition that has been deemed harmless, most famously by Lewis, at (1974a), 66–7). Nevertheless, for Quine, counterfactual conditionals do have a vital scientific use in connection with dispositions – at least those expressing dispositions narrowly so-called are relatively immune to changes in context, and are often indispensable to science. Solubility in water, for example, can be expressed as 'if x were immersed in water, x would dissolve'. Such forms are not extensional, and therefore according to Quine must be regarded as second best, not to be used in a reality-limning, regimented statement of a theory (1960, 222–5). But a disposition, for Quine, is in turn identical with an underlying structural trait which can, in principle, be expressed in extensional language as a universal generalisation – even if we cannot, given the present state of knowledge, specify which (1974a, 8–15; 1995a, 21; in the case of solubility, the relevant generalisation pertains to chemical structure).

The key is the role of universal generalisations, not exactly dispositions as such. The idea is that legitimate science has no 'ungrounded' true counterfactual conditionals, none that cannot in principle be explained in terms of extensional, universal generalisations. True, this is still not as rich an apparatus as the full gamut of possibility made available on Williamson's conception, but I put it that it is sufficient for such uses as Galileo's in his thought experiment.

13.3 Conclusion

Williamson is not alone in assuming that certain notions of everyday life – knowledge, belief, reference, assertion, causality, essence, possibility, conceivability, and so on – are fit notions in terms of which to frame philosophical questions and answers. Mustering a genuinely persuasive argument against Williamson's instinct would be extremely difficult, and would amount to something on the order of Kant's *Critique of Pure Reason* or Wittgenstein's *Investigations*, revised and augmented. But what I want to point out, and emphasise, is that neither Quine nor Carnap make such an assumption. In fact, their not making it is the motive arguably most distinctive and central to their philosophies. In a word, they regard almost every such notion as standing at least in need of what they call explication, and many of them, for Quine most visibly with his penchant for desert landscapes but Carnap too, represent curiosities which are perhaps natural but have no place in science.[11]

---

[11] Work on this essay was supported by a visiting professorship provided by the Philosophical Faculty of the University of Hradec Králové in Summer Term 2018.

Bibliography

Adler, Mortimer. (1941). "God and the Professors" Presentation at Conference on Science, Philosophy, and Religion. www.ditext.com/adler/gp.html
Alspector-Kelly, Marc. (2001). "On Quine on Carnap on Ontology." *Philosophical Studies* 102:1, 93–122.
Awodey, Steve and Carsten Klein, eds. (2004). *Carnap Brought Home*. Chicago: Open Court.
Ayer, A. J., ed. (1959). *Logical Positivism*. Glencoe, IL: Free Press.
Baghramian, M. and S. Marchetti, eds. (2018). *Pragmatism and the European Traditions*. London: Routledge.
Bain, A. (1859). *The Emotions and the Will*. London: J.W. Parker & Sons.
Becker, Edward. (2012). *The Themes of Quine's Philosophy: Meaning, Reference, and Knowledge*. Cambridge: Cambridge University Press.
Ben-Menahem, Y. (2006). *Conventionalism*. Cambridge: Cambridge University Press.
⸺ (2016). "The Web and the Tree: Quine and James on the Growth of Knowledge." In Kemp and Janssen-Lauret, 59–75.
Benson, Arthur. (1963). "Bibliography of Rudolf Carnap." In Schilpp, 1015–1070.
Bird, Graham H. (1995). "Internal and External Questions." *Erkenntnis* 42:1, 41–64.
Bird, Alexander. (2012). "*The Structure of Scientific Revolutions* and Its Significance: An Essay Review of the Fiftieth Anniversary Edition." *British Journal for the Philosophy of Science* 63:4, 859–883.
Blatti, Stephan and Sandra Lapointe, eds. (2016). *Ontology after Carnap*. Oxford: Oxford University Press.
Blumberg, A. and H. Feigl. (1931). "Logical Positivism: A New Movement in European Philosophy." *The Journal of Philosophy* 28:11, 281–296.
Bonhert, Herbert G. (1963). "Carnap's Theory of Definition and Analyticity." In Schilpp, 407–430.
Boolos, G., J. Burgess, and R. Jeffrey, eds. (2002). *Computability and Logic*, 4th ed. Cambridge, Cambridge University Press.
Borradori, Giovanna. (1994). *The American Philosopher*. Trans. Rosanna Crocitto. Chicago: Chicago University Press.
Brower, R. (1959). *On Translation*. Cambridge, MA: Harvard University Press.
Carnap, Rudolf. (1922a). *Der Raum*, Kant-Studien 56. Berlin: Reuther & Reichard.

(1922b). "Vom Chaos zur Wirklichkeit." Ms. in Carnap-Nachlass RC 081-05-01, Archive of Scientific Philosophy, Hillman Library, University of Pittsburgh.
(1923). "Über die Aufgabe der Physik und die Anwendung des Grundsatzes der Einfachstheit." *Kant-Studien* 28:1/2, 90–107.
(1924). "Dreidimensionalität des Raumes und Kausalität." *Annalen der Naturphilosophie und philosophischen Kritik* 4:1, 105–130.
(1926). *Physikalische Begrifsbildung*. Karlsruhe: Braun.
(1927). "Eigentliche und uneingentliche Begriffe." *Philosophische Zeitschrift für Forschung und Aussprache* 1:4, 355–374.
(1928a). *Der logische Aufbau der Welt*. Berlin: Weltkreis.
(1928b). Scheinprobleme in der Philosophie, Berlin: Bernary, trans. as "Pseudoproblems in Philosophy." In The Logical Structure of the World World/ Pseudoproblems in Philosophy. Berkeley: University of California Press, 1967, repr. Chicago: Open Court, 2003.
(1928/1967). *The Logical Structure of the World and Pseudoproblems in Philosophy*. Trans. Rolf George. Berkeley: University of California Press.
(1929). Abriß der Logistik. Mit besonderer Ber ücksichtigung der Relationstheorie und ihrer Anwendungen. Vienna: Springer.
(1930). "Die Mathematik als Zweig der Logik." *Blätter für Deutsche Philosophie* 4:3/4, 298–340.
(1932a). "Die physikalische Sprache als Unversalsprache der Wissenschaft." *Erkenntnis* 2:1, 432–465. Trans. *The Unity of Science*, London: Kegan, Paul, Trench Teubner & Co., 1934.
(1932b). "Psychologie in physikalischer Sprache." *Erkenntnis* 3:1, 107–142. Trans. "Psychology in Physicalist Language." In Ayer, 165–198.
(1932c). "Über Protokollsätze." *Erkenntnis* 3:1, 204–214.
(1932/1959). "The Elimination of Metaphysics Through the Logical Analysis of Language." Reprinted in Ayer, 60–81.
(1932/1987). "On Protocol Sentences." Trans. by R. Creath and R. Nollan. *Noûs* 21:4, 457–470.
(1934a). *Logische Syntax der Sprache*. Vienna: Springer.
(1934b). "Meaning, Assertion and Proposal." *Philosophy of Science* 1:3, 359–360.
(1934c). "On the Character of Philosophical Problems." *Philosophy of Science* 1:1, 5–19.
(1934d). "The Task of the Logic of Science." Trans. Hans Kaal. In McGuinness, 46–66.
(1934/1937). *The Logical Syntax of Language*. Trans. Amethe Smeaton. London: Routledge and Kegan Paul Ltd.
(1935). *Philosophy and Logical Syntax*. Bristol: Thoemmes Press, 1996.
(1936a). "Testability and Meaning." *Philosophy of Science* 3:4, 420–471.
(1936b). "Von der Erkenntnistheorie zur Wissenschaftslogik." In *Actes du Congrès Internationale de Philosophie Scientifique*, Paris: Herman and Cie, 36–51.
(1937). "Testability and Meaning – Continued." *Philosophy of Science* 4:1, 1–40.

(1939). "Foundations of Logic and Mathematics." In *Encyclopedia of Unified Science*, eds. Otto Neurath, Rudolf Carnap, and Charles W. Morris, Vol. 1, No. 3, Chicago: University of Chicago Press, 139–213.
(1942). *Introduction to Semantics*. Cambridge, MA: Harvard University Press.
(1946). "Remarks on Induction and Truth." *Philosophy and Phenomenological Research* 6:4, 590–602.
(1947). *Meaning and Necessity*. Chicago: University of Chicago Press.
(1950a). "Empiricism, Semantics, and Ontology." *Revue Internationale de Philosophie* 4:11, 20–40.
(1950b). *Logical Foundations of Probability*. Chicago: University of Chicago Press.
(1950/1956). "Empiricism, Semantics, and Ontology." Reprinted in his (1956), 205–221.
(1952). "Meaning Postulates." *Philosophical Studies* 3:5 65–73. Reprinted as Appendix B in the second edition of *Meaning and Necessity* (University of Chicago Press, 1956).
(1952/1990). "Quine on Analyticity." Written in 1952; first published in Quine and Carnap (1990), 427–432.
(1955). "Meaning and Synonymy in Natural Languages." *Philosophical Studies* 6:3, 33–47.
(1956). *Meaning and Necessity: A Study in Semantics and Modal Logic*. 2nd enlarged ed. Chicago: University of Chicago Press.
(1958a). "Beobachtungssprache und theoretische sprache." *Dialectica* 12:3–4, 236–248. Translated and reprinted as "Observation Language and Theoretical Language." In Hintikka, 75–85.
(1958b). *Introduction to Symbolic Logic and Its Applications*. Toronto, Canada: Dover Publications, Inc.
(1963a). "Intellectual Autobiography." In Schilpp, 1–89.
(1963b). "Replies and Systematic Expositions." In Schilpp, 859–1016.
(1963c). "Wilfrid Sellars on Abstract Entities in Semantics." In Schilpp, 923–927.
(1963d). "W. V. Quine on Logical Truth." In Schilpp, 915–922.
Carroll, Lewis. (1895). "What the Tortoise Said to Achilles." *Mind* 4:14, 278–280.
Carus, André. (2007a). *Carnap and Twentieth-Century Thought. Explication as Enlightenment*. Cambridge: Cambridge University Press.
(2007b). "Carnap's Intellectual Development." In Friedman and Creath, 19–42.
(2016). "Carnap and Phenomenology: What happened in 1924?" in *Influences on the Aufbau*, ed. Christian Damböck (Vienna: Springer), 137–162.
Chalmers, David, David Manley, and Ryan Wasserman, eds. (2009). *Metametaphysics*. Oxford: Clarendon Press.
Chang, Hasok. (2008). "Contingent Transcendental Arguments for Metaphysical Principles." *Royal Institute of Philosophy* Supplement 63: 113–133.
Chomsky, Noam. (2000). *New Horizons in the Study of Language and Mind*. Cambridge: Cambridge University Press.

Cook, G. A. (2013). "Resolving Two Key Problems in Mead's *Mind, Self, and Society*." In *George Herbert Mead in the 21st Century*, eds. F. Thomas Burke and Krzysztof Piotr Skowronski. New York: Lexington Press, 95–106.
Costreie, Sorin, ed. (2016). *Early Analytic Philosophy – New Perspectives on the Tradition*. Cham, Switzerland: Springer.
Creath, Richard. (1990a). "Introduction." In Quine and Carnap (1990), 1–43.
  (1990b). "Quine, Carnap, and the Rejection of Intuition." In *Perspectives on Quine*, eds. Roger B. Barrett and Roger F. Gibson. Cambridge, MA: Blackwell, 55–66.
  (1990c). "The Unimportance of Semantics." *PSA: Proceedings of the Biennial Meeting of the Philosophy of Science Association* 1990:2, 405–416.
  (1991). "Every Dogma Has Its Day." *Erkenntnis* 35:1/3, 347–389.
  (1994). "Functionalist Theories of Meaning and the Defense of Analyticity." In Salmon and Wolters, 287–304.
  (2004). "Quine on the Intelligibility and Relevance of Analyticity." In Gibson, 47–64.
  (2007). "Quine's Challenge to Carnap." In Friedman and Creath, 316–335.
  (2009). "The Gentle Strength of Tolerance: The Logical Syntax of Language and Carnap's Philosophical Programme." In Wagner (2009), 203–214.
  (2012a). "Analyticity in the Theoretical Language." In Creath (2012), 57–66.
  (2012b). "Before Explication." In Wagner (2012), 161–174.
Creath, Richard, ed. (2012). *Rudolf Carnap and the Legacy of Logical Empiricism*. Dordrecht: Springer.
Creath, Richard and Michael Friedman, eds. (2007). *The Cambridge Companion to Carnap*. New York: Cambridge University Press.
Curd, Martin and Stathis Psillos, eds. (2008). *The Routledge Companion to Philosophy of Science*. New York: Routledge.
David, Marian. (1996). "Analyticity, Carnap, Quine, and Truth." *Philosophical Perspectives* 10, 281–296.
Davidson, Donald. (1973). "Radical Interpretation." Reprinted in *The Essential Davidson*. New York: Oxford University Press, 2006, 184–195.
  (1975). "Thought and Talk." Reprinted in his *Inquiries into Truth and Interpretation*. Oxford: Clarendon Press, 1984, 155–170.
  (1986). "A Nice Derangement of Epitaphs." Reprinted in *The Essential Davidson*, ed. E. Lepore and K. Ludwig. Oxford: Clarendon Press, 251–265.
  (1992). "The Second Person." Reprinted in *Subjective, Intersubjective, Objective*. Oxford: Oxford University Press, 2001a, 107–122.
  (2001a). "Comments on the Karlovy Vary Papers." In Kotatko et al., 285–308.
  (2001b). "Externalisms." In Kotatko et al., 1–16.
Davidson, Donald and Jaakko Hintikka, eds. (1969). *Words and Objections: Essays on the Work of W. V. Quine*. Dordrecht: D. Reidel Publishing Company.
Demos, Raphael. (1939). *The Philosophy of Plato*. New York: Charles Scribner's Sons.
  (1948). "Note on Plato's Theory of Ideas." *Philosophy and Phenomenological Research* 8:3, 456–460.

Devitt, Michael. (2008). "Realism/Anti-Realism." In Curd and Psillos, 256–267.
Devlin, William J. and Alisa Bokulich, eds. (2015). *Kuhn's Structure of Scientific Revolutions – 50 Years On*. New York: Springer.
De Waal, Cornelis. (2008). "A Pragmatist World View: George Herbert Mead's Philosophy of the Act." In *The Oxford Handbook of American Philosophy*, ed. Cheryl Misak. Oxford: Oxford University Press, 144–168.
Domski, Mary and Michael Dickson Dickson, eds. (2010). *Discourse on a New Method*. Chicago: Open Court.
Dreben, Burton. (1992). "Putnam, Quine – and the Facts." *Philosophical Topics* 20:1. The Philosophy of Hilary Putnam. 293–315.
Ebbs, Gary. (1997). *Rule-Following and Realism*. Cambridge, MA: Harvard University Press.
  (2009). *Truth and Words*. Cambridge: Cambridge University Press.
  (2011). "Carnap and Quine on Logical Truth." *Mind* 120:478 (April 2011), 193–237.
  (2017). *Carnap, Quine, and Putnam on Methods of Inquiry*. New York: Cambridge University Press.
  (2019). "Carnap on Analyticity and Existence: A Clarification, Defense, and Development of Quine's Reading of Carnap's Views on Ontology." *Journal for the History of Analytical Philosophy* 7:5, 1–31. https://jhaponline.org/jhap/article/view/3876
Edmister, Bradley and Michael O'Shea. (1994). "W.V. Quine: Perspective on Logic, Science, and Philosophy." Interview. Reprinted in Quine (2008b), 43–56.
Eklund, Matti. (2009). "Carnap and Ontological Pluralism." In Chalmers, Manley, and Wasserman, 130–156.
  (2013). "Carnap's Metaontology." *Noûs* 47:2, 229–249.
Enderton, Herbert. (1972). *A Mathematical Introduction to Logic*. New York: Academic Press, Inc.
Faris, Ellsworth. (1936). "Review of *Mind, Self, and Society*." *American Journal of Sociology* 41:6, 809–813.
Feigl, Herbert. (1950). "The Mind-Body Problem in the Development of Logical Empiricism." Reprinted in Feigl, *Inquiries and Provocations*, ed. R. S. Cohen. Dordrecht: Reidel, 1981, 286–301.
Feigl, H. (1975). "Homage to Rudolf Carnap." In Hintikka, xiii–xvii.
Feigl, Herbert and Grover Maxwell, eds. (1962). *Scientific Explanation, Space, and Time*. Vol. 3. *Minnesota Studies in the Philosophy of Science*. Minneapolis: University of Minnesota Press.
Floyd, Juliet and Sanford Shieh, eds. (2001). *Future Pasts: The Analytic Tradition in Twentieth-Century Philosophy*. New York: Oxford University Press,
Fodor, Jerry and Ernest Lepore. (1992). *Holism: A Shopper's Guide*. Oxford: Blackwell.
Franco, Paul. (2020). "Hans Reichenbach's and C. I. Lewis's Kantian Philosophies of Science." *Studies in History and Philosophy of Science Part A* 80: 62–71.
Frank, Philipp. (1963). "The Pragmatic Components in Carnap's 'Elimination of Metaphysics.'" In Schilpp, 159–164.

Friedman, Michael. (1987). "Carnap's *Aufbau* Reconsidered." Reprinted in Friedman (1999), 89–113.
   (1992). "Epistemology in the *Aufbau*." Reprinted with postscript in Friedman (1999), 114–164.
   (1999). *Reconsidering Logical Positivism*. Cambridge: Cambridge University Press.
   (2003). "Kuhn and Logical Empiricism." In Nickles, 19–44.
   (2007). "The *Aufbau* and the Rejection of Metaphysics." In Creath and Friedman, 129–153.
   (2010). "Synthetic History Reconsidered." In Domski and Dickson, 573–813.
   (2012). "Kuhn and Philosophy." *Modern Intellectual History* 9:1, 77–88.
Frost-Arnold, Greg. (2011). "Quine's Evolution from 'Carnap's Disciple' to the Author of 'Two Dogmas'." *HOPOS: Journal of the International Society for the History of Philosophy of Science* 1:2, 291–316.
   (2013). *Carnap, Tarski, and Quine at Harvard: Conversations on Logic, Mathematics, and Science*. Chicago: Open Court.
George, Alexander. (2000). "On Washing the Fur without Wetting It: Quine, Carnap, and Analyticity." *Mind*, New Series, 109:433, 1–24.
Gibson, Roger F., ed. (2004). *The Cambridge Companion to Quine*. New York: Cambridge University Press.
Glock, Hans-Johann. (2003). *Quine and Davidson on Language, Thought and Reality*. Cambridge: Cambridge University Press.
Grice, H. P. and P. F. Strawson. (1956). "In Defense of a Dogma." *The Philosophical Review* 65:2, 141–158.
Gupta, Anil. (2006). *Empiricism and Experience*. Oxford: Oxford University Press.
Gustafsson, Martin. (2014). "Quine's Conception of Explication – and Why It Isn't Carnap's." In Harman and Lepore (2014), 508–525.
Hacker, P. M. S. (1996). *Wittgenstein's Place in Twentieth-Century Analytic Philosophy*. Oxford: Blackwell.
Hahn, L. E. and P. A. Schilpp, eds. (1986). *The Philosophy of W.V. Quine*. La Salle, Il: Open Court.
   (1998). *The Philosophy of W.V. Quine*. Expanded ed. La Salle, IL: Open Court.
Hardcastle, Gary and Alan Richardson, eds. (2003). *Logical Empiricism in North America*. Minneapolis: University of Minnesota Press.
Harman, Gilbert and Ernie Lepore, eds. (2014). *A Companion to W.V.O. Quine*. Malden, MA: Wiley-Blackwell.
Heidelberger, Michael. (2003). "The Mind-Body Problem in the Origin of Logical Empiricism: Herbert Feigl and Psychophysical Parallelism." In *Logical Empiricism. Historical and Contemporary Perspectives*, ed. Paolo Parrini, Wesley Salmon, and Merrilee Salmon. Pittsburgh: University of Pittsburgh Press, 233–262.
Hempel, Carl G. (1950). "Problems and Changes in the Empiricist Criterion of Meaning." *Revue international de philosophie* 41:11, 41–63.
   (1951). "The Concept of Cognitive Significance: A Reconsideration." *Proceedings of the American Academy of Arts and Sciences* 80:1, 61–77.
Hilbert, David. (1899). *Grundlagen der Geometrie*. Leipzig: Teubner.

Hillier, Sam. (2010). "Analyticity and Language Engineering: Carnap's Logical Syntax." *European Journal of Analytic Philosophy* 6:2, 25–46.

Hintikka, Jaakko, ed. (1975). *Rudolf Carnap: Logical Empiricist*. Synthese Library 73. Dordrecht: Reidel.

Hookway, Christopher. (2008). "Pragmatism and the Given: C. I. Lewis, Quine, and Peirce." In *The Oxford Handbook of American Philosophy*, ed. Cheryl Misak. Oxford: Oxford University Press, 269–289.

Horwich, Paul, ed. (1993). *World Changes: Thomas Kuhn and the Nature of Science*. Cambridge, MA: MIT Press.

Howard, Don. (2003). "Two Left Turns Make a Right: On the Curious Political Career of North American Philosophy of Science." In Hardcastle and Richardson, 25–93.

  (2018). "Quine, Dewey, and the Pragmatist Tradition in American Philosophy of Science." Lecture at HOPOS conference Groningen, Holland.

Huebner, Daniel. (2012). "The Construction of *Mind, Self, and Society*: The Social Process Behind G. H. Mead's Social Psychology." *Journal of the History of the Behavioral Sciences* 48:2, 134–153.

Hylton, Peter. (2001). "'The Defensible Province of Philosophy': Quine's 1934 Lectures on Carnap." In Floyd and Shieh, 257–276.

  (2004). "Quine on Reference and Ontology." In Gibson, 115–150.

  (2007). *Quine*. New York: Routledge.

  (2013). "Quine and the *Aufbau*: The Possibility of Objective Knowledge." In Reck, 78–92.

  (2014a). "Quine's Naturalism Revisited." In Harman and Lepore, 148–62.

  (2014b). "Significance in Quine." *Grazer Philosophische Studien* 89, 113–133.

  (2021). "Carnap and Quine on Analyticity: The Nature of the Disagreement." *Noûs* 55:2, 445–462 .

  (forthcoming). "Moorean Propositions and Russellian Confusion." In *Early Analytic Philosophy: Origins and Transformations*, ed. James Conant and Gilad Nir. New York: Routledge.

Jacquette, Dale. (2003). *Philosophy, Psychology, and Psychologism: Critical and Historical Readings on the Psychological Turn in Philosophy*. New York: Kluwer Academic.

James, William. (1890). *The Principles of Psychology*. New York: Holt.

  (1907/1909/1955). *Pragmatism and Four Essays from The Meaning of Truth*. New York: Meridian Books.

  (1897/1956). *The Will to Believe and Other Essays in Popular Philosophy*. New York: Dover.

Järvilehto, Lauri. (2009). "The Pragmatic A Priori of C. I. Lewis." *Cognitio-Estudos* 6:2, 96–102.

Joas, Hans. (1985). *G. H. Mead: A Contemporary Re-Examination of His Thought*. Cambridge, MA: MIT Press.

Kemp, Gary. (2014). "Quine's Relationship with Analytic Philosophy." In Harman and Lepore, 69–88.

  (2016). "Underdetermination, Realism, and Transcendental Metaphysics in Quine." In Kemp and Janssen-Lauret, 168–188.

Kemp, Gary and Frederique Janssen-Lauret, eds. (2016). *Quine and His Place in History*. New York: Palgrave Macmillan.
Kim, Jaegwon. (2003). "Logical Positivism and the Mind-Body Problem." In *Logical Empiricism: Historical and Contemporary Perspectives*. Ed. P. Parrini, M. Salmon, and W. Salmon. Pittsburgh: University of Pittsburgh Press, 263–279.
Klagge, J. and A. Nordmann, eds. (1993). *Ludwig Wittgenstein: Philosophical Occasions*. Indianapolis: Hackett.
Klein, Alexander. (2008). "Divide Et Impera! William James's Pragmatist Tradition in the Philosophy of Science." *Philosophical Topics* 36:1, 129–166.
Kotatko, P., P. Pagin, and G. Segal, eds. (2001). *Interpreting Davidson*. Stanford: CSLI.
Kripke, Saul. (1975). "Outline of a Theory of Truth." *Journal of Philosophy* 72:19, 690–716.
  (1982). *Wittgenstein on Rules and Private Language*. Cambridge, MA: Harvard University Press.
Kuhn, Thomas. (1962/1970). *The Structure of Scientific Revolutions*. 3rd ed. Chicago: University of Chicago Press, 1996.
  (1970). "Reflections on My Critics." In Lakatos and Musgrave, 231–278.
  (1990). "The Road Since Structure." Reprinted in his (2000), 90–104.
  (1993). "Afterwards." In Horwich, 311–341.
  (2000). *The Road Since Structure: Philosophical Essays, 1970–1993, with an Autobiographical Interview*. Ed. James Conant and John Haugel. Chicago: University of Chicago Press.
Lakatos, Imre and Alan Musgrave, eds. (1970). *Criticism and the Growth of Knowledge*. New York: Cambridge University Press.
Leonardi, Paolo and Marco Santambrogio, eds. (1995). *On Quine: New Essays*. Cambridge: Cambridge University Press.
Lepore, Ernest and Kirk Ludwig. (2005). *Davidson: Meaning, Truth, Language and Reality*. Oxford: Oxford University Press.
Lewis, C. I. (1929). *Mind and the World-Order*. New York: Charles Scribner's Sons.
  (1946). *An Analysis of Knowledge and Valuation*. La Salle, IL: Open Court.
Lewis, David. (1973). *Counterfactuals*. Oxford: Blackwell.
Levine, James. (2009). "From Moore to Peano to Watson: The Mathematical Roots of Russell's Naturalism and Behaviorism." *The Baltic International Yearbook of Cognition, Logic and Communication* 4, 1–126.
  (2016). "The Place of Vagueness in Russell's Philosophical Development." In Costreie, 161–212.
Limbeck-Lilienau, Christoph. (2012) "Carnap's Encounter with Pragmatism." In Creath (2012), 89–112.
Livingston, Paul. (2008). *Philosophy and the Vision of Language*. New York: Routledge.
MacBride, Fraser. (2021). "Rudolf Carnap and David Lewis on Metaphysics: A Question of Historical Ancestry." *Journal for the History of Analytical Philosophy* 9:1, 1–31.

Manninen, Juha. (2002). "Wie entstand der Physikalismus?" *Nachrichten. Forschungsstelle und Dokumentationszentrum für Österreichische Philosophie, Graz* 10, 22–52.
  (2003). "Towards a Physicalistic Attitude." In *The Vienna Circle and Logical Empiricism. Re-evaluation and Future Perspectives.* Ed. F. Stadler. Dordrecht: Kluwer, 133–150.
Martin, R. M. (1952). "On 'Analytic'," *Philosophical Studies* 3(3): 42–47.
Massey, G. J. (1978). "Indeterminacy, Inscrutability, and Ontological Relativity." *American Philosophical Quarterly, Monograph* 12, 43–55.
Massimi, Michela. (2015). "Walking the Line: Kuhn Between Realism and Relativism." In Devlin and Bokulich, 135–152.
Maxwell, Grover. (1962). "The Ontological Status of Theoretical Entities." In Feigl and Maxwell, 3–15.
McGuinness, Brian, ed. (1987). *Unified Science: The Vienna Circle Monographs Series Originally.* Ed. Otto Neurath. Dordrecht: D. Reidel.
Mead, G. H. (1934). *Mind, Self, & Society: From the Standpoint of a Social Behaviorist.* Ed. Charles Morris. Chicago: Chicago University Press.
  (1964). *Selected Writings.* Ed. Andrew Reck. Chicago: Chicago University Press.
Mills, Charles W. (2007). "White Ignorance." In *Race and Epistemologies of Ignorance*, Ed. Shannon Sullivan and Nancy Tuana. New York: Suny Press, 13–38.
Misak, Cheryl. (2013). *The American Pragmatists.* Oxford: Oxford University Press.
Moore, G. E. (1993). "Wittgenstein's Lectures in 1930–33." In Klagge and Nordmann, 46–114.
Mormann, Thomas. (2007). "Carnap's Logical Empiricism, Values, and American Pragmatism." *Journal for General Philosophy of Science* 38:1, 127–146.
  (2010). "History of Philosophy of Science As Philosophy of Science by Other Means?" In Stadler, 29–40.
  (2012a). "A Place for Pragmatism in the Dynamics of Reason?" *Studies in the History and Philosophy of Science Part A* 43, 27–37.
  (2012b). "Toward a Theory of the Pragmatic A Priori." In Creath, 113–132.
  (2016). "Morris' Pariser Programm einer wissenschaftlichen Philosophie." In *Wissenschaft und Praxis.* Ed. Christian Bonnet and Elisabeth Nemeth. Vienna: Springer, 73–88.
Morris, Charles. (1925). *Symbolism and Reality: A Study in the Nature of Mind.* Philadelphia: John Benjamins Publishing Company.
  (1934). "Introduction." In Mead, v–xxxv.
  (1936). "The Concept of Meaning in Pragmatism and Logical Positivism." *Actes du huitieme congres international de philosophie.* 2–7 September, Prague, 130–138.
  (1937). *Logical Positivism, Pragmatism, and Scientific Empiricism.* Paris: Hermann et Cie.
  (1937/1979). *Logical Positivism, Pragmatism, and Scientific Empiricism.* Providence, RI: American Mathematical Society, reprint of 1937 Hermann edition.
  (1938). *Foundations of the Theory of Signs. International Encyclopedia of Unified Science*, 1, 2, Chicago: Chicago University Press, 1938. Reprinted in Neurath et al. (1969), 78–137.

(1942). "Empiricism, Religion, and Democracy." In *Science, Philosophy, and Religion: Second Symposium*. Ed. Lyman Bryson and Louis Finkelstein. New York: Conference on Science, Philosophy and Religion in their Relation to the Democratic Society, 213–241.

(1946). *Signs, Language, and Behavior*. New York: George Braziller.

(1963). "Pragmatism and Logical Empiricism." In Schilpp, 87–98.

Mulvaney, Robert J. and Philip M. Zeltner, eds. (1981). *Pragmatism: Its Sources and Prospects*. Columbia: University of South Carolina Press.

Murphey, Murray. (2012). *The Development of Quine's Philosophy*. Boston: Springer Science.

Neurath, Otto. (1910). "Zur Theorie der Sozialwissenschaften." *Jahrbuch für Gesetzgebung, Verwaltung und Volkswirtschaft im Deutschen Reich* 34, 37–67. Trans. "On the Theory of Social Science" in Neurath, *Economic Writings: Selections 1904–1945* (ed. by T. Uebel and R. S. Cohen), Dordrecht: Kluwer, 2004, 265–291.

(1928). "Rezension: R. Carnap, *Der Logische Aufbau der Welt* und *Scheinprobleme der Philosophie*." *Der Kampf* 21, 624–626. Reprinted in Neurath, *Gesammelte philosophische und methodologische Schriften* (ed. by R. Haller and H. Rutte), Vienna: Hölder-Pichler-Tempsky, 1981, 295–297.

(1931). "Physikalismus." *Scientia* 50, 297–303. Trans. "Physicalism" in Neurath (1983), 52–57.

(1932). "Protokollsätze." *Erkenntnis* 3:1, 204–214. Trans. "Protocol Sentences" in Ayer 1959, 199–208, and "Protocol Statements" in Neurath (1983), 91–99.

Neurath, Otto, Rudolf Carnap, and Charles Morris, eds. (1969). *Foundations of the Unity of Science: Toward an International Encyclopedia of Unified Science*, Vol 1, Nos 1–10. Chicago: Chicago University Press.

Nickles, Thomas, ed. (2003). *Thomas Kuhn*. New York: Cambridge University Press.

O'Shea, James. (2018). "The Analytic Pragmatist Conception of the A Priori: C. I. Lewis and Wilfrid Sellars." In Baghramian and Marchetti, 203–227.

Orenstein, Alex and Petr Kotatko, eds. (2000). *Knowledge, Language and Logic: Questions for Quine*. Dordrecht: Kluwer.

Pearson, James. (2011). "Distinguishing W.V. Quine and Donald Davidson." *Journal for the History of Analytical Philosophy* 1:1, 1–22.

(2017). "Taking Care with Quine's 'Don't-Cares.'" *The Monist* 100:2, 266–287.

Peirce, C. S. (1868). "Some Consequences of Four Incapacities." *Journal of Speculative Philosophy* 2:3, 140–157.

(1877/1966). "The Fixation of Belief." In *Selected Writings*. Ed. P. P. Wiener. New York: Dover, 91–112.

(1878/1966) "How to Make Our Ideas Clear." In *Selected Writings*. Ed. P. P. Wiener. New York: Dover, 113–136.

Pincock, Christopher. (2005). "A Reserved Reading of Carnap's *Aufbau*." *Pacific Philosophical Quarterly* 86:4, 518–543.

Planck, Max. (1950). *Scientific Autobiography and Other Papers*. Trans. Frank Gaynor. London: Williams and Norgate.

Price, Huw. (2007). "Quining Naturalism." *Journal of Philosophy* 104:8, 375–402.

(2009). "Metaphysics after Carnap." Reprinted in Price (2011), 280–303.
(2011). *Naturalism Without Mirrors*. New York: Oxford University Press.
Psillos, Stathis and Martin Curd, eds. (2008). *The Routledge Companion to Philosophy of Science*. New York: Routledge.
Putnam, Hilary. (1962). "The Analytic and the Synthetic." Reprinted in his (1975), 33–69.
(1975). *Mind, Language, and Reality*. New York: Cambridge University Press.
(1976). "Realism and Reason." Reprinted in his (1978), 123–138.
(1978). *Meaning and the Moral Sciences*. Boston: Routledge and Kegan Paul Ltd.
Quine, W. V. (1932). "The Logic of Sequences. A Generalization of *Principia Mathematica*." Ph.D. Dissertation, Harvard University.
(1934). "Lectures on Carnap." In Quine and Carnap (1990), 47–103.
(1935). "Review of *Logische Syntax der Sprache* by Rudolf Carnap." *Philosophical Review* 44:4, 394–397.
(1936). "Truth by Convention." Reprinted in Quine (1976), 70–99.
(1937). "Is Logic a Matter of Words?" (unpublished manuscript), abstract in *Journal of Philosophy* 34:1937, 674.
(1939). "Designation and Existence." *The Journal of Philosophy* 36:26, 701–709.
(1940). *Mathematical Logic*. Cambridge, MA: Harvard University Press.
(1943). "Notes on Existence and Necessity." *Journal of Philosophy* 40:5, 113–127.
(1948a). "On What There Is." *The Review of Metaphysics* 2:5, 21–38.
(1948b). "On What There Is." Reprinted in Quine (1980), 1–19.
(1949). "Animadversions on the Notion of Meaning." In Quine (2008a), 152–156.
(1950). *Methods of Logic*. New York: Holt.
(1950/1980). "Identity, Ostension, and Hypostasis." Reprinted in Quine (1980), 65–79.
(1951a). "On Carnap's Views on Ontology." Reprinted in Quine (1976), 203–211.
(1951b). "On Carnap's Views on Ontology." *Philosophical Studies* 2:5, 65–72.
(1951c). "The Present State of Empiricism." Appendix 6 to Verhaegh (2018). Transcription by S. Verhaegh.
(1951d). "The Problem of Meaning in Linguistics." In Quine (1980), 47–64.
(1951e). "Two Dogmas of Empiricism." *Philosophical Review* 60:1, 20–43.
(1951f). "Two Dogmas of Empiricism." Reprinted in Quine (1980), 20–46.
(1953a). "Meaning and Existential Inference." In Quine (1980), 160–167.
(1953b). "Notes on the Theory of Reference." Reprinted in Quine (1980), 130–138.
(1957). "The Scope and Language of Science." Reprinted in his (1976), 228–245.
(1955). "Posits and Reality." Reprinted in his (1976), 246–254.
(1958). "Le mythe de la signification." In *La Philosophie Analytique*. Ed. Cahiers de Royaumont, 4. Paris: Minuit, 1962, 139–169.
(1959). "Meaning and Translation." In Brower, 148–172.
(1960). *Word and Object*. Cambridge, MA: MIT Press.
(1963a). "Carnap and Logical Truth." In Schilpp, 385–406.
(1963b). "Necessary Truth." Reprinted in Quine (1976), 68–76.

(1969a). "Epistemology Naturalized." In Quine (1969a), 69–90.
(1969b). "Natural Kinds." In Quine (1969a), 114–138.
(1969c). "Ontological Relativity." In Quine (1969d), 26–68.
(1969d). *Ontological Relativity and Other Essays*. New York: Columbia University Press.
(1969e). "Replies." In Davidson and Hintikka, 292–352.
(1969f). "Reply to Chomsky." In Davidson and Hintikka, 302–311.
(1970a). "Homage to Carnap." Reprinted in Quine (1976), 40–43.
(1970b). "On the Reasons for Indeterminacy of Translation." Reprinted in Quine (2008a), 209–214.
(1970c). "On the Reasons for the Indeterminacy of Translation." *The Journal of Philosophy* 67:16, 178–183.
(1970d). *Philosophy of Logic*. London: Prentice-Hall.
(1974a). *Roots of Reference*. LaSalle, IL: Open Court Publications.
(1974b). "Skinner's Retirement Party." In Føllesdal and Quine (2008b), 291–292.
(1975). "The Pragmatists' Place in Empiricism." In Mulvaney and Zeltner, 21–39.
(1976). *The Ways of Paradox and Other Essays*. Rev. and enlarged ed. Cambridge, MA: Harvard University Press.
(1979). "Cognitive Meaning." In Quine (2008a), 288–302.
(1980). *From a Logical Point of View: Nine Logico-Philosophical Essays*. 2nd rev. ed., Cambridge, MA: Harvard University Press.
(1981a) "Five Milestones of Empiricism." In Quine (1981c), 67–72.
(1981b). "Responding to Schuldenfrei." In Quine (1981c), 184–186.
(1981c). *Theories and Things*. Cambridge, MA: Harvard University Press, 1981.
(1981d). "Things and Their Place in Theories." In Quine (1981c), 1–23.
(1984). "Relativism and Absolutism." Reprinted in Quine (2008a), 319–322.
(1985). *The Time of My Life: An Autobiography*. Cambridge, MA: MIT Press.
(1986a). "Autobiography." In Hahn and Schilpp, 3–46.
(1986b). *Philosophy of Logic*, 2nd ed. Cambridge, Mass.: Harvard University Press.
(1987). "Indeterminacy of Translation Again." Reprinted in Quine (2008a), 341–346.
(1989). *Quiddities: An Intermittently Philosophical Dictionary*. Cambridge, MA: Harvard University Press.
(1990a). "Comment on Creath." In Barrett and Gibson, 67.
(1990b). "Comments." In Barrett and Gibson.
(1990c). *Pursuit of Truth*. Cambridge, MA: Harvard University Press.
(1990d). "Three Indeterminacies." In Barrett and Gibson, 1–16.
(1991). "Two Dogmas in Retrospect." Reprinted in Quine (2008a), 390–400.
(1992a). *Pursuit of Truth*. Revised edition. Cambridge, MA: Harvard University Press.
(1992b). "Structure and Nature." Reprinted in Quine (2008a), 401–406.
(1994a). "Comment." Reprinted as "Comments on Neil Tennant's 'Carnap and Quine'" in Dagfinn Follesdal and Douglas B. Quine (2008b), 216–222.

(1994b). "Exchange between Donald Davidson and W. V. Quine Following Davidson's Lecture." In his (2008b), 152–156.

(1994c). "Indeterminacy Without Tears." Reprinted in Quine (2008a), 447–448.

(1995a). *From Stimulus to Science*. Cambridge, MA: Harvard University Press.

(1995b). "Reactions." In Leonardi and Santambrogio, 347–361.

(2000). "Responses." In Orenstein and Kotatko, 407–430.

(2008a). *Confessions of a Confirmed Extensionalist and Other Essays*. Ed. Dagfinn Føllesdal and Douglas B. Quine. Cambridge, MA: Harvard University Press.

(2008b). *Quine in Dialogue*. Ed. Dagfinn Føllesdal and Douglas B. Quine. Cambridge, MA: Harvard University Press.

Quine, W. V. and Rudolf Carnap. (1990). *Dear Carnap, Dear Van*. Ed. Richard Creath. Berkeley: University of California Press.

Correspondence. Quine Archive, Houghton Library, MS 2587, item 197, box 7.

Quine, W. V. and Charles Morris. Correspondence. Quine Archive, Houghton Library, MS 2587 item 741, box 27; item 547, box 20.

Quine, W. V. and J. S. Ullian. (1978). *The Web of Belief*, 2nd ed. New York: McGraw-Hill, Inc.

Rawling, Piers and Philip and Wilson, eds. (2019). *The Routledge Handbook of Translation and Philosophy*. New York: Routledge.

Reck, Erich, ed. (2013). *The Historical Turn in Analytic Philosophy*. New York: Palgrave Macmillan.

Reisch, George. (1991). "Did Kuhn Kill Logical Positivism?" *Philosophy of Science* 58:2, 264–277.

(2005). *How The Cold War Transformed Philosophy of Science*. Cambridge: Cambridge University Press.

Rhees, Rush. (1970). *Discussions of Wittgenstein*. London: Routledge and Kegan Paul.

Richardson, Alan. (1994). "Carnap's Principle of Tolerance." *Proceedings of the Aristotelian Society* 68: 67–83.

(1997). "Two Dogmas about Logical Empiricism: Carnap and Quine on Logic, Epistemology, and Empiricism." *Philosophical Topics* 25:2, 145–168.

(1998). *Carnap's Construction of the World: The Aufbau and the Emergence of Logical Empiricism*. Cambridge: Cambridge University Press.

(2003). "Logical Empiricism, American Pragmatism, and the Fate of Scientific Philosophy in North America." In Hardcastle and Richardson, 1–24.

(2007). "That Sort of Everyday Image of Logical Empiricism: Thomas Kuhn and the Decline of Logical Empiricist Philosophy of Science." In Richardson and Uebel, 346–370.

Richardson, Alan and Thomas Uebel, eds. (2007). *The Cambridge Companion to Logical Empiricism*. New York: Cambridge University Press.

Ricketts, Thomas. (1982). "Rationality, Translation, and Epistemology Naturalized." *Journal of Philosophy* 79:3, 117–136.

(2004). "Frege, Carnap, and Quine: Continuities, and Discontinuities." In Awodey and Klein, 181–202.

(2010). "Quine's Objection and Carnap's Aufbau." In Domski and Dickson, 313–331.
Rorty, R. (1997). "Introduction." In Sellars, iii–vi.
Rosen, Gideon. (2014). "Quine and the Revival of Metaphysics." In Harman and Lepore, 552–70.
Roth, Paul A. (1978). "Paradox and Indeterminacy." *Journal of Philosophy* 75:7, 347–367.
  (1984). "On Missing Neurath's Boat." *Synthese* 61:2, 205–231.
  (1999). "The Epistemology of 'Epistemology Naturalized.'" *Dialectica* 53:2, 87–109.
  (2003). "Why There is Nothing Rather than Something: Quine on Behaviorism, Meaning, and Indeterminacy." In Jacquette, 263–287.
  (2006). "Naturalism Without Fears." In Turner and Risjord, 684–708.
  (2008). "Epistemology of Science after Quine." In Psillos and Curd, 3–14.
  (2019). "Quine." In Rawling and Wilson, 104–121.
  (2020). *The Philosophical Structure of Historical Explanation*. Evanston: Northwestern University Press.
Russell, Bertrand. (1903/1937). *The Principles of Mathematics*, 2nd ed. London: George Allen and Unwin.
  (1912/1999). *The Problems of Philosophy*. Minneola, NY: Dover.
  (1914/1993). *Our Knowledge of the External World As a Field for Scientific Method in Philosophy*. New York: Routledge.
Ryckman, Thomas. (2007). "Carnap and Husserl." In Friedman and Creath, 81–105.
Sacks, Mark. (2000). *Objectivity and Insight*. Oxford: Clarendon Press.
Salmon, W. and G. Wolters, eds. (1994). *Language, Logic, and the Structure of Scientific Theories*. Pittsburgh: University of Pittsburgh Press.
Schilpp, P. A., ed. (1963). *The Philosophy of Rudolf Carnap, The Library of Living Philosophers*. La Salle, IL.: Open Court.
Schlauch, Margaret. (1947). "The Cult of the Proper Word." *New Masses* 63:3, 15–18.
Sellars, Wilfrid. (1956). "Empiricism and the Philosophy of Mind." In *The Foundations of Science and the Concept of Psychology and Psychoanalysis* (Minnesota Studies in the Philosophy of Science, Vol. 1). Ed. H. Feigl and M. Scriven. Minneapolis: University of Minnesota Press, 253–329. Reprinted as Sellars, *Empiricism and the Philosophy of Mind* (ed. R. Brandom), Cambridge, MA: Harvard University Press, 1997.
Sharrock, Wes and Rupert Read. (2002). *Kuhn: Philosopher of Scientific Revolution*. Cambridge: Polity Press.
Shapin, Steven. (2015). "Kuhn's *Structure*: A Moment in Modern Naturalism." In Devlin and Bokulich, 11–22.
Sheldon, William. (1954). *Atlas of Men*. New York: Harper.
Sinclair, Robert. (2012). "Quine and Conceptual Pragmatism." *Transactions of the Charles S. Peirce Society* 48:3, 335–355.

(2016). "On Quine's Debt to Pragmatism: C. I. Lewis and the Pragmatic A Priori." In *Quine and His Place in History*, ed. G. Kemp and F. Janssen-Lauret. Basingstoke: Palgrave-Macmillan, 76–99.
Smart, J.C.C. (1969). "Quine's Philosophy of Science." In Davidson and Hintikka, 3–13.
Soames, Scott. (2003). *Philosophical Analysis in the Twentieth Century*, Volume I. Princeton: Princeton University Press.
　(2009). "Ontology, Analyticity, and Meaning: The Quine–Carnap Dispute." In Chalmers, Manley, and Wasserman, 424–443.
Stadler, Friedrich, ed. (2010). *The Present Situation in the Philosophy of Science*. Dordrecht: Springer.
Stein, Howard. (1992). "Was Carnap Entirely Wrong, After All?" *Synthese* 93:½, 275–295.
Stich, Stephen and Kevin Tobia. (2018). "Intuition and Its Critics." In Stuart, Fehige, and Brown, 369–384.
Strawson, P. F. (1950). "On Referring." *Mind* 59:235, 320–344
Stuart, Michael T., Yiftach Fehige, and James Robert Brown, eds. (2017). *The Routledge Companion to Thought Experiments*. London: Routledge.
Stump, David. (2015). "Alternative Conceptions of the A Priori: Cassirer, Lewis, and Pap." In his *From Conceptual Change and the Philosophy of Science: Alternative Interpretations of the A Priori*. New York: Routledge, 90–118.
Tarski, Alfred. (1936). "The Concept of Truth in Formalized Languages." In Tarski (1983), 152–277.
　(1944). "The Semantic Conception of Truth: and the Foundations of Semantics." *Philosophy and Phenomenological Research* 4:3, 341–376.
　(1983). *Logic, Semantics, Meta-Mathematics: Papers from 1923 to 1938*, 2nd ed., trans. J. H. Woodger, ed. John Corcoran. Indianapolis: Hackett.
Tennant, Neil. (1994). "Carnap and Quine." In Salmon and Wolters, 315–344
Thomasson, Amie. (2016). "Carnap and the Prospects for Easy Ontology." In *Ontology after Carnap*, ed. Stephan Blatti and Sandra Lapointe. Oxford: Oxford University Press, 122–144.
Tsou, Jonathan. (2015). "Reconsidering the Carnap–Kuhn Connection." In Devlin and Bokulich, 51–70.
Turner, Stephen and Mark Risjord, eds. (2006). *Philosophy of Anthropology and Sociology*. New York: Elsevier.
Uebel, Thomas. (2001). "Carnap and Neurath in Exile: Can Their Disputes Be Resolved?" *International Studies in the Philosophy of Science* 15:2, 211–220.
　(2004). "Carnap, the Left Vienna Circle, and Neopositivist Antimetaphysics." In Awodey and Klein, 247–277.
　(2007a). *Empiricism at the Crossroads: The Vienna Circle's Protocol Sentence Debate*. Chicago: Open Court.
　(2007b). "Carnap and the Vienna Circle: Rational Restructionism Refined." In Friedman and Creath, 129–152.
　(2010). "Some Remarks on Current History of Analytical Philosophy of Science." In Stadler (2010), 13–28.

(2013). "Pragmatics in Carnap and Morris and the Bipartite Metatheory Conception." *Erkenntnis* 78:3, 523–546.

(2014). "Carnap's Aufbau and Physicalism: What does the 'Mutual Reducibility' of Psychological and Physical Objects Amount to?" In *European Philosophy of Science – Philosophy of Science in Europe and the Viennese Heritage*. Ed. M. C. Galavotti, E. Nemeth, and F. Stadler. Dordrecht: Kluwer, 45–56.

(2016). "Neurath's Influence on Carnap's *Aufbau*." In *Influences on the Aufbau*, ed. C. Damböck, Cham: Springer, 51–76.

(2021). "Rejecting the Given: Neurath and Carnap on Methodological Solipsism." *HOPOS* 11:2.

van Fraassen, Bas. (1980). *The Scientific Image*. Oxford: Clarendon Press, 1980.

Verhaegh, Sander. (2017). "Boarding Neurath's Boat: The Early Development of Quine's Naturalism." *Journal of the History of Philosophy* 55:2, 317–342.

(2018). *Working From Within: The Nature and Development of Quine's Naturalism*. New York: Oxford University Press.

(2019a). "The Behaviorisms of Skinner and Quine: Genesis, Development, and Mutual Influence." *Journal of the History of Philosophy* 57:4, 707–730.

(2019b). "Sign and Object: Quine's Forgotten Book Project." *Synthese* 196:12, 5038–5060.

(2020a). "The American Reception of Logical Positivism: First Encounters (1929–1932)." *HOPOS: The Journal of the International Society for the History of Philosophy of Science* 10:1, 106–142.

(2020b). "Coming to America: Carnap, Reichenbach, and the Great Intellectual Migration. Part I: Rudolf Carnap." *Journal for the History of Analytical Philosophy* 8:11, 1–23.

(2020c). "Coming to America: Carnap, Reichenbach, and the Great Intellectual Migration. Part II: Hans Reichenbach." *Journal for the History of Analytical Philosophy* 8:11, 24–47.

Von Foerster, Heinz and Bernhard Poerksen. (2002). *Understanding Systems: Conversations on Epistemology and Ethics*. Dordrecht: Springer.

Wagner, Pierre, ed. (2009). *Carnap's Logical Syntax of Language*. New York: Palgrave Macmillan.

(2012). *Carnap's Ideal of Explication and Naturalism*. New York: Palgrave Macmillan.

White, Morton. (1950). "The Analytic and the Synthetic: an Untenable Dualism." In *John Dewey: Philosophy of Science and Freedom*. Ed. Sidney Hook. New York: The Dial Press, 316–330.

(1999). *A Philosopher's Story*. University Park: Penn State University Press.

Williamson, Timothy. (2007). *The Philosophy of Philosophy*. Oxford: Blackwell.

(2014). "How Did We Get Here from There? The Transformation of Analytic Philosophy." *Belgrade Philosophical Annual* 27, 7–37.

(2015). *Modal Logic as Metaphysics*. Oxford: Oxford University Press.

Wittgenstein, Ludwig. (1922/1990). *Tractatus Logico-philosophicus*. Trans. C. K. Ogden. London: Routledge and Kegan Paul.

(1958). *Philosophical Investigations*, 2nd ed. Oxford: Blackwell.

(1969a). *The Blue and Brown Books*, 2nd ed. Oxford: Blackwell.

(1969b). *On Certainty*. Oxford: Blackwell.
(1974). *Philosophical Grammar*. Oxford: Blackwell.
(1975). *Philosophical Remarks*. Oxford, Blackwell.
(1977). *Remarks on Colour*. Oxford: Blackwell.
(1979). *Wittgenstein's Lectures 1932–1935*. Chicago: University of Chicago Press.
(1998). *Culture and Value*, revised ed. Oxford: Blackwell.
Yablo, Stephen. (1998). "Does Ontology Rest on a Mistake?" *Proceedings of the Aristotelian Society*. Supplementary Volume 72: 229–261.
Zilsel, Edgar. (1932). "Bemerkungen zur Wissenschaftslogik." *Erkenntnis* 3, 143–161.

Index

analytic/synthetic distinction, 27, 89, 102–103, 116–118, 121, 122, 128–131, 135–137, 149–151, 155, 161, 164, 166, 207–208, 227–230, 232–233, 238–239, 240, 244–245, 250–251, 257–259, 262
analytic, trivially, 227–232

Bentham, Jeremy, 88

Carnap, Ina, 11, 23
Carnap, Rudolf, 18, 68, 93, 108, 196, 210, 218–219, 261
  Abriss der Logistik, 13
  analytic, definition of, 142–145
  analytic/synthetic distinction, 116–117, 121, 130, 136–137
  Aufbau, discussions with Vienna Circle, 35
  Aufbau, intertranslatability in, 40–45
  Aufbau, structuralism of, 34, 36
  categoricity, 79
  conventionalism, 78, 87
  Der Logische Aufbau der Welt, 18–19, 33, 86, 99
  Der Raum, 79
  "Eigentliche und uneigentliche Begriffe," 79–81
  "Empiricism, Semantics, and Abstract Objects," 54–57
  epistemology, 28, 220, 222–223, 225, 226, 249
  explication, 193, 208, 233, 262
  formal languages, 256
  foundations of mathematics, debates in, 81
  indeterminacy of translation, 192–193
  intensional and extensional languages, 119
  intension/extension distinction, 169–172
  linguistic frameworks, 54–57, 68, 216–219, 222, 225, 242
  logic of science, 29, 193, 253
  Logical Syntax of Language, 23–25, 54, 99
  Löwenheim-Skolem theorem, 79
  "Meaning and Synonymy in Natural Languages," 118–119, 169–173
  meaning, theory of, 169–175
  metaphysics, 24, 53–55, 192, 219–220, 224, 228, 233, 235–236, 243
  methodological solipsism of, 33, 37, 38
  methodological solipsism, abandonment of, 49
  Morris, response to, 112
  move to the United States, 86
  objectivity, 99, 102
  ontological commitment, 216
  ontology, 54–57, 235–236, 241, 243, 257
  philosophy and science, 193
  physicalism, 39, 40, 42, 44–46
  pragmatism, 74
    as epistemic discretion, 74, 81–82
    of distinguished from Quine, 84
    American, 74–75, 86–87
  rational reconstruction, 259–260
  Schlick, and, 35
  semantical rules, use of, 151
  semantics, pure and descriptive, 139–140
  synthetic a priori, 192
  "Testability and Meaning," 49
  translation, 140
  "Über Protokollsätze," 19–20
  United States, move to the, 30–31
Carroll, Lewis, 156, 163
category/subclass distinction, 226, 228–232
Chomsky, Noam, 269
conceptual engineering. *See* explication
Cooley, John, 18–19, 28, 29
Creath, Richard, 116, 120

Davidson, Donald, 92, 116–118
  epistemology, 106–107
  Quine, critique of, 107–109
Day, Doris, 194
Demos, Raphael, 126
Descartes, Rene, 75, 106

Index

Dewey, John, 88, 104
Dollfuss, Engelbert, 22
Dreben, Burton, 69

Ebbs, Gary, 120, 161–163, 216, 227–228, 231
Einstein, Albert, 184, 189
empiricism, 89
empiricism, British, 94
Encyclopedia for Unified Science, 93
explication, 148–149, 163, 255, 260, 263–267, 269
external/internal distinction, 220

Feigl, Herbert, 13–14
Frank, Philipp, 101
Frege, Gottlob, 80, 88

Galileo, Galilei, 188, 266, 267, 269
Gettier, Edmund, 266–268

Hacker, P.M.S., 179
Heidegger, Martin, 259
Hempel, Carl G., 158
Hilbert, David, 79, 80
Hitler, Adolf, 22, 112
Howard, Don, 88
Huntington, E.V., 115
Hylton, Peter, 111, 159, 258

indeterminacy of translation, 181–182
    Wittgenstein, Ludwig, 179, 182
instrumentalism, 199
intensional, 130–131
internal questions and external questions, 166
internal/external distinction, 82, 217, 220–222, 227–233, 240–245, 256–257
    Carnap on, 54–57
    Quine against, 85

James, William, 75, 76, 78, 88–90, 114, 128–129
Joas, Hans, 95

Kripke, Saul, 179
Kuhn, Thomas
    naturalism, 68
    ontology, 68
    paradigms and world changes, 63–65
    real world, 68
    realism, 62, 65
    reality not independent of theory, 67–68
    relativism and scientific progress, 65–67
    science as about the world, 63
    science as puzzle solving, 66–67
    science, progress of, 65–68
    scientific communities, 65
    stimuli and sensations, 64–65

Lewis, Clarence Irving, 12, 13, 26, 86, 88, 115, 130–131
    An Analysis of Knowledge and Valuation, 124–126
    conceptual pragmatism, 114, 130
    interpretation, 127
    meaning, 126
    Mind and the World Order, 127
    Quine, influence on, 15, 25
Limbeck-Lilienau, Christoph, 86
Livingston, Paul, 175
Locke, John, 88
logical positivism, 88, 94, 97, 111
logocentric predicament, 156

Martin, Richard M., 137
Mead, George Herbert, 95–98, 100, 110
meaning, theory of, 206
meaning, verifiability criterion of, 155
meaningfulness, 154
metaphysics, 104, 119–120, 224–225
Morris, Charles, 31, 86, 92–94, 108, 111–113, 157
    analytic/synthetic distinction, 102
    Carnap, criticisms of, 98–102
    meaning, 157–158
    ontology, 103–104
    protocol sentence debate, 98
    semiotics, 97–98
    verification, 157
Murphey, Murray, 123

Nagel, Ernest, 86, 111
Nazism, 22, 86
Neurath, Otto, 20, 33, 93, 104
    Aufbau, early criticism of, 37–40
    Aufbau, review of, 36–37
    Carnap, influence on, 46–48
    Der Logische Aufbau der Welt, 46–49
    physicalism of, 49
nominalism, 225

objectivity
    socialized, 109–111
    ontology, 233–234, 244–245, 251–252

Peano, Giuseppe, 80
Peirce, Charles Sanders, 75, 77, 88, 90, 114, 127–130
Plato, 126
Platonism, 224
Poincaré, Henri, 79

pragmatic
  Carnap and Quine appeals to, 253
pragmatism
  as epistemic discretion, 74
pragmatism, American, 94
  characteristics of, 75–78

Quine, Naomi, 14, 21, 23
Quine, Willard Van Orman, 15–16, 58–59, 65,
    68, 86, 248–249, 261
  analyticity, 238–239
  and ontology, 165
  contrasted with truth, 141
  analytic/synthetic distinction, 27, 85, 207
  Aufbau, reading of, 18
  behaviorism, 27–28, 179
  Carnap, first meeting with, 21
  Carnap, lectures on (1934), 29–31
  Carnap, Prague meeting with, 22–23
  Carnap, Rudolf, first meeting with, 14
  category/subclass distinction, 220–222,
    226–227, 232–233
  conceptual pragmatism, critique of, 128–130
  convention, 162–163
  conventionalism, 25
  counterfactual conditionals, 270
  Der Logische Aufbau der Welt, 28, 32–35
  dissertation, publication of, 15, 17, 21
  empiricism, 121
  epistemology, 25, 197, 208–210, 249–250
  epistemology naturalized, 108
  extensionalism, 196, 208
  fallibilism, 61
  formal languages, 255
  holism, 15, 91, 209
  indeterminacy of reference, 206, 208, 237–238
  indeterminacy of translation, 159–160, 174,
    177–178, 181–182, 205
    and inscrutability of reference, 83
    and ontology, 83
    and underdetermination of theory, 82
  instrumentalism, 61
  intensional/extensional distinction, 114–115,
    118–120, 122–123, 130
  intensionality, 196, 206
  language as social, 180
  language, acquisition of, 190
  language, conception of, 202
  Lewis, critical of, 115, 123–124
  logic, 188, 210
  logic and mathematics, 25
  Logical Syntax of Language, 25–27, 29
  Löwenheim-Skolem theorem, 83
  mathematics, 188
  maxim of minimum mutilation, 74

  meaning, 84, 159, 174, 185–187, 190, 193
    as use, 181
    theory of, 207
  metaphysics, 61, 102, 164, 169, 180, 224, 236
  Morris, response to, 112
  naturalism, 59–61, 253
  naturalized epistemology, 107
  objectivity, 106
  observation, 199
  observation sentences, 105–107
  ontological commitment, 216
  ontological relativity, 208
  ontology, 60–61, 104, 165–166, 196, 208,
    236–237, 239, 259
  phenomenalism, 16, 29
  philosophy as syntax, 26
  philosophy, method of, 188–189
  physicalism, 50–51, 199
  Poland, visit to, 29
  pragmatic considerations, appeals to, 73–74
  pragmatism, 60, 130
    of distinguished from American
      pragmatism, 74
    of distinguished from Carnap, 84
    American, 87–88
  "The Pragmatists' Place in Empiricism," 87–88
  private language, 180–181
  "The Problem of Meaning in Linguistics," 167–168
  psychology, 198
  radical translation, 178, 183, 203
  radical translation as thought experiment, 183
  realism, 57, 59–61, 199, 206
  reference, 196, 208
  reference, theory of, 207
  science and philosophy, 189–190
  science, demarcation of, 162
  semantical rule, criticism of, 149–151
  semantical rules, criticism of, 145–148
  Sheldon Fellowship, 12–14, 17
  simplicity, 198–200, 204, 206, 209
  synonymy, 201, 206, 208, 263
  synthetic a priori, against, 187–188
  theory of meaning not explanatory, 160, 163–164
  translation, 182–183
  truth, 84, 90, 91, 174–175, 196
  "Truth by Convention," 28, 156
  "Two Dogmas of Empiricism," 194
  underdetermination, 73, 197–201, 205–206
  Vienna, first trip to, 17–18
  "On What There Is," 58–59
  Word and Object, 200–207

realism, metaphysical, 53
Reichenbach, Hans, 86
Richardson, Alan, 103, 120

Royce, Josiah, 114
Russell, Bertrand, 80, 88, 247–248
  Our Knowledge of the External World, 18, 28, 32
  theory of types, 223–224

Sacks, Mark, 109
scepticism, 76
Schlick, Moritz, 17, 18, 35, 38
science
  Carnap and Quine's sense of, 254
scientific philosophy, 253
semantical rules, 151–153
Sheffer, H.M., 115
Sheffer, Henry, 12–13
Sinclair, Robert, 123
Strawson, Peter, 264

Tarski, Alfred, 137–141, 151
tolerance, principle of, 20, 24, 28, 29, 54, 78, 81, 85, 100, 101, 120, 122–123, 240, 246–252, 257, 258
Tooke, John Horne, 88
truth, 77–78, 137–142, 151

underdetermination, 90
Unity of Science movement, 93, 97

verification, 157, 158
verification theory of meaning, 164
Vienna Circle, 13, 18, 20–21, 157

White, Morton, 115, 121–122, 126, 128, 130
Whitehead, Alfred North, 12, 15, 103, 115
Williamson, Timothy, 268–269
  analyticity, 262
  philosophical questions, 261
  semantics, 263–264
  synonymy, 263
  thought experiments, 265–267
  vagueness, 264
Wittgenstein, Ludwig, 24, 88
  behaviorism, 179
  logic, 188
  mathematics, 188
  meanings, 185–186
  philosophy, method of, 188–189
  private language, 184
  radical translation, 178
  science and philosophy, 189–190
  synthetic a priori, against, 187–188
  Tractatus Logico-Philosophicus, 13
  translation, 182–183

Printed by Printforce, United Kingdom